Energy Audits

Energy Audits

The Key to Delivering Real Energy Reductions

By Kit Oung with Steven Fawkes and John Mulholland

bsi.

First published in the UK in 2014

By
BSI Standards Limited
389 Chiswick High Road
London W4 4AL

©The British Standards Institution 2014

All rights reserved. Except as permitted under the Copyright, Designs and Patents Act 1988, no part of this publication may be reproduced, stored in a retrieval system or transmitted in any form or by any means – electronic, photocopying, recording or otherwise – without prior permission in writing from the publisher.

Whilst every care has been taken in developing and compiling this publication, BSI accepts no liability for any loss or damage caused, arising directly or indirectly in connection with reliance on its contents except to the extent that such liability may not be excluded in law.

While every effort has been made to trace all copyright holders, anyone claiming copyright should get in touch with the BSI at the above address.

BSI has no responsibility for the persistence or accuracy of URLs for external or third-party internet websites referred to in this book, and does not guarantee that any content on such websites is, or will remain, accurate or appropriate.

The right of Kit Oung, Steven Fawkes and John Mulholland to be identified as the authors of this work has been asserted by them in accordance with Sections 77 and 78 of the Copyright, Designs and Patents Act 1988.

Typeset in Great Britain by Letterpart Limited, www.letterpart.com

Printed in Great Britain by Berforts Group, www.berforts.co.uk

British Library Cataloguing in Publication Data

A catalogue record for this book is available from the British Library

ISBN 978-0-580-82247-6

Contents

About the authors	1
Introduction	3
Chapter 1 Managing energy and auditing	7
The twin functions of energy management	7
The role of top management	16
ISO 50001, *Energy management systems — Requirements with guidance for use*	17
The uses of energy audit in energy management	22
Chapter 2 Energy audit standards	24
Standardizing energy audits	27
Different uses of energy audit standards	33
What does an energy audit not solve?	43
Chapter 3 Understanding energy use, energy consumption and energy efficiency	45
Of boilers, cooling towers, chillers, air compressors and ventilation systems	46
A machine can be energy efficient and still use more energy	52
Identifying energy use and variables that cause consumption to vary	54
Energy performance indicators and energy baselines	65
Modern day thinking about energy reduction	69
Chapter 4 Defining the boundaries of an energy audit	78
Energy Efficiency Directive (EED)	78
Applicability of the EED in the UK	79
Implications of ESOS on energy audits	82
Scoping the energy audit	89
Why large enterprises only? Does it mean SMEs will not achieve any benefits?	98
Reporting: Creating transparency for energy initiatives	99
Chapter 5 The processes of an energy audit	105
Planning the energy audit	106
Roles and responsibilities	109
Opening meeting	110
Data and information	111

Observing the organizational activities and physical operating
    conditions 115
Analysing opportunities for improvement 116
Energy audit reporting 122
Closing meeting 124

Chapter 6 Using energy audit processes to maximize energy savings 125

Chapter 7 Choosing the right team 154
Hiring preferences 155
Energy audit competence 158
Associated skills 161
Likeability 164
Other 'small print' 165
Buyers beware 168

Chapter 8 Implementing energy reduction 170
It is about management not technology 170
Project identification, development and evaluation 172
Technical appraisal 173
Contextual appraisal 173
Economic evaluation 174
Reducing uncertainty 175
Financing energy performance improvement investments 176
Project management 177
Post-project analysis 177

Appendix A People aspects/behavioural change 179
Identifying the current situation 180
Employee environmental awareness surveys 180
Campaign strategy plan 182
Maintaining momentum 187

Appendix B List of complementary standards 188

Further Reading 196

References 197

About the authors

Kit Oung

Kit Oung BEng MSc(Eng) CEng MIChemE MEI MCMI is a practising energy consultant specializing in energy and carbon reduction strategies using low-cost high-return opportunities and energy management systems. He has 16 years' experience in energy auditing and implements energy reduction projects for blue-chip, multinational companies, which includes covering petrochemical, speciality chemical, pharmaceutical, food and beverage and large commercial properties across five continents.

Kit is a recognized expert and he regularly reviews and provides inputs to various energy management and environmental management standards. He was the convenor for EN 16247-3, *Energy audits – Part 3: Processes*, project leader for ISO 50002, *Energy audits – Requirements with guidance for use* and the technical author for the Publicly Available Specification (PAS) 51215, *Energy efficiency assessment – Competence of a lead energy assessor – Specification*.

Kit frequently writes about business sustainability, energy management and energy efficiency, including two books: *Energy Management in Business: The Manager's Guide to Maximising and Sustaining Energy Reduction* (Gower, 2013) and *Implementing and Improving an Energy Management System: How to Meet the Requirements of ISO 50001* (BIP 2221:2013) (BSI, 2013).

Steven Fawkes

Steven Fawkes BSc DipTechEcon PhD PCSB CEng FEI FBIS is an internationally recognized expert on energy efficiency with more than 30 years' experience, including implementing energy management programmes for corporates, and local and national governments, and co-founding two energy service companies, one in Romania and one in the UK, that implemented innovative energy services contracts with Diageo and Sainsbury's.

From 2007 to 2012 he was a partner at Matrix, where he led the Extel number one-rated research team and advised clean tech companies on fundraising and M&A. He has been an adviser to the UK Department of Energy & Climate Change and in 2012 was awarded the Energy Institute's Individual Achievement Award. He has published extensively on energy efficiency, including two books published by Gower. He is currently a

About the authors

director of, or senior adviser to, several energy and clean tech companies and funds, and he is implementing efficiency finance programmes in Europe and North America.

John Mulholland

Eur Ing John Mulholland BScTech (Hons) CEng CSci MIChemE MEI is Director of Mulholland Energy Solutions, which specializes in reducing energy consumption by behavioural interventions in large organizations.

John has worked in the energy sector for 40 years and holds a degree in chemical engineering and fuel technology. For 15 years he worked as a chemical engineer reducing energy consumption in process plants and for 24 years he was with NIFES Consulting Group, holding the position of Director of People and Management Solutions. John has worked in 12 countries for large organizations in industry, commerce and the public sector. He is currently writing a book for Gower called *Greening the Workforce: Energy Programmes and Employee Behaviour*.

Introduction

On a global basis, Planet Earth has an estimated 1,600 gigatons [1] of carbon available as sources of energy such as coal, Liquefied Petroleum Gas (LPG), diesel, heavy fuel oil and natural gas. These fossil fuels took Mother Nature millions of years to create. When these fossil fuels are consumed, they are converted into carbon dioxide and emitted into the atmosphere.

When occurring naturally, Planet Earth absorbs carbon dioxide from the atmosphere at a rate of two gigatons per year. Before the Industrial Revolution, the rate of carbon consumption was roughly equal to the rate at which carbon dioxide was adsorbed from the atmosphere. However, since the Industrial Revolution, the rate of carbon consumption has risen dramatically.

In the 1920s, the rate of energy consumption was approximately one gigaton of carbon per year. By the 1950s this had doubled. In 2006, the figure had risen to eight gigatons of carbon per year. McKinsey & Company predicts that global energy demand is likely to grow at a rate of 2.2 per cent until 2020.[2] Fifty-nine per cent of the extracted carbon is consumed as fuel sources to generate heat and power. Present projections indicate that there will be a shortfall of oil in the latter half of the twenty-first century.

As 'easily' extractable sources of energy are depleted, the technology needed to extract the 'more difficult and costly' sources of energy will be required. In a separate study, McKinsey & Company [3] reported that the average cost of bringing new oil wells online has risen by 100 per cent over the past decade. Apart from nuclear energy, substituting fossil fuel energy with renewable forms is difficult because there is insufficient land available for food and living spaces and to install solar panels, wind turbines and/or crops for fuel.[4]

The balance of probability is that, in a world where demand of energy outweighs its supply, the cost of energy will continue to rise as the scarcity of energy sources increases. Traditionally, when a company faces economic hardship, many companies choose to lay off employees, treating energy as a fixed cost for the organization. Bain & Company [5] found that this practice is diminishing: more and more companies are realizing that, while generating very short-term benefits, nearly 60 per

Introduction

cent of downsizing, outsourcing and business process re-engineering exercises are failing to regain business profitability.

Fifty-nine per cent of the carbon in the atmosphere comes from the process of burning fossil fuel to generate heat and power. A by-product of combustion is the generation of carbon dioxide, CO_2. Seventeen per cent of carbon in the atmosphere is a by-product of deforestation and a subsequent reduction in Planet Earth's ability to sequester CO_2. Fourteen per cent of carbon in the atmosphere comes naturally from agriculture and livestock. The remaining comes from other greenhouse gases (GHGs).[6]

The net result from an increase in carbon consumption from energy use outweighing the rate at which carbon is naturally sequestered is that CO_2 accumulates in the atmosphere, building up in concentration and giving rise to the climate change phenomenon and its mitigation. Climate change debates and controversies have centred on the consequences of increasing CO_2 concentrations in the atmosphere, the prediction of dangerous levels of CO_2 and the timeline to reach this critical limit.

The debates were not about the fact that CO_2 is building up in the atmosphere nor about CO_2 being a contributor to climate change. In fact, multidisciplinary research led by Johan Rockström found climate change to be one of nine human activities putting the planet at risk from irreversible change. The others are: rate of biodiversity loss, interference with nitrogen and phosphorous cycles, stratospheric ozone depletion, ozone acidification, fresh water use, change in land use, chemical pollution and atmospheric aerosol loading.[7] Climate change from energy consumption is also one of the easier aspect to address.

The UK finds itself in a unique position in that, due to a lack of large quantities of private investment in low-carbon power generation and power plant closures due to end-of life assets, Ofgem (Office of Gas and Electricity Markets) predicts that the excess capacity for the UK is in the region of 2 to 5 per cent.[8] That is to say, assuming there are no natural, climatic changes and/or catastrophic failures in the pipeline, and we have 2 to 3 per cent excess capacity over the next few years, there is no risk of blackouts and grey outs. If any of these incidents occurs, the risk of blackouts and grey outs increases.

Fraunhofer Institute for Systems and Innovation Research in Germany [9] recommends that Europe has a potential to reduce its energy consumption by 57 per cent. The building stock could see a 71 per cent reduction through better insulation, modern construction techniques and energy-efficient ventilation, heating and cooling. In an industrial setting, this could be as high as 52 per cent and the transport sectors could achieve a 53 per cent reduction via better traffic management and logistics.

In a study by McKinsey & Company, up to 25 per cent of these energy savings do not require major capital costs or involve significant changes in business processes.[10] As such, opportunities to save energy are real and achievable. Implementing energy savings not only results in immediate financial savings for the organization, but also has wider political, economic, social and environmental benefits.

In fact, businesses are beginning to become aware of the competitive benefits of energy reduction: the direct cost reduction, a reduction in associated losses and waste (e.g. maintenance, water, effluent and waste), improved cost accounting, lower-cost options for future expansion, maximizing the profit margin, a high return on investment, attracting top talent [11] and motivating staff, attracting investors [12], brand reputation, gaining market share and profiting from being green.[13]

Some organizations are beginning to look beyond the traditional short-term financial gains and compliance, to long-term risk management [14] and strategic importance. The 2012 Edelman goodpurpose® study found that more than 70 per cent surveyed said they would recommend, promote and switch brands to those with good environmental and sustainability performance.[15] In fact, Generation Ys (those born in either the 1980s or the 1990s) are 90 per cent more likely to want to be working for and/or consuming products and services from companies with good environmental and sustainability track records.[16]

This has led to a mushrooming of product and service offerings devoted to energy reduction: energy auditing, energy studies, energy management, energy management audits, energy reviews, energy surveys, energy diagnostics, etc. Within the many naming conventions, there are many different scopes of works (or supplies), degrees of thoroughness and, to some extent, degrees of software automation. All of these messages can be confusing and seemingly disjointed...at least for the layperson who needs to manage energy consumption and energy costs.

For this reason, the international community has developed a management systems standard for managing energy (ISO 50001) and energy auditing standards (the EN 16247 series and ISO 50002).

One hundred and thirty-six ethnographic studies [17] found that people, naturally and socially, do not use the terms 'energy conservation' and 'energy efficiency'. They readily identify, however, with the terms 'energy savings' or 'energy reduction'. The study also found that people associate 'energy efficiency' with new machines or equipment they purchase. Yet, replacing a still-functioning machine or equipment for a more energy-efficient model is thought to be 'wasteful'.

Introduction

This book, written for business managers, business owners, entrepreneurs and energy managers, is a companion to ISO 50002 but mirrors the colloquial speech of saving energy or reducing energy waste in small- to medium-sized enterprises (SMEs) and in large organizations. It focuses on energy auditing as a tool to identify opportunities to save energy, and its links with energy management and the Energy Efficiency Directive (EED).

Chapter 1 and Chapter 2 put the role of energy auditing into the context of organizations' endeavours to manage energy consumption, and why organizations carry out an energy audit, and provide a short background on energy auditing standards in Europe and internationally.

Chapter 3 introduces the concepts of energy use, energy consumption and energy efficiency. It highlights areas where energy information and energy-related information can be obtained and gives an introduction to how they can be used to generate an energy baseline and energy performance indicators.

Chapter 4 introduces the requirements of the EED and the UK's interpretation: the Energy Savings Opportunity Scheme (ESOS). It also covers a framework that can be used to define a scope and boundaries that meet the regulatory requirements.

Chapter 5 describes the processes of an energy audit and highlights the requirements placed on the energy auditor and the organization. When carrying out an energy audit, there are activities that an organization can do to facilitate the energy auditor and there are ways to make the energy audit output insightful and valuable for the organization.

Chapter 6 uses the energy maturity matrix to describe how various opportunities for improvement can be stacked up into a portfolio that maximizes energy reduction and minimizes capital cost.

Chapter 7 describes the importance of how choosing to work with a competent person adds value to the organization. It gives a simple framework, consistent with PAS 51215, for identifying and shortlisting such a competent person.

Finally, Chapter 8 introduces the often neglected step – to turn the output from an energy audit into real savings: financing and implementing energy reduction projects.

The engineering and scientific calculations have been purposefully left out from the scope of this book. Should you find an interest or need to look at the engineering details of energy reduction, please refer to any good energy engineering books available on the market.

Good luck in your journey.

Chapter 1   Managing energy and auditing

Creating sustainable models, 'greening' the boardroom, and applying disruptive innovations that help organizations manage the risk of energy prices have gained much ground in recent years. There are a lot of renewable technologies and 'low-energy' technologies available to support companies and public bodies to become greener. These are all good opportunities to reduce energy consumption and many companies have seized the opportunity to invest in them.

A wise manager can use these technologies in a portfolio to create, maximize and sustain low-cost, high-return energy reduction, and minimize the organizational risks at the same time. It requires the organization to reign in and manage energy as part of its operations. Logically, managers need to do two things: first, identify and implement opportunities to reduce energy consumption and then, secondly, implement and improve on existing governance in order to sustain or maintain the reduced energy consumption within the organization. These are the twin functions of energy management.

The twin functions of energy management

Energy reduction – building up energy maturity

A significant majority of companies do not know where they use energy and treat it as a fixed-cost component in their operations. The first step for managers in reducing energy consumption is to know where the organization uses energy. Then, find out how much energy the organization should be using to deliver business benefits.

In an office building, energy is used to supply fresh air for the occupants and to extract the stale air. Energy is also required to condition the fresh air: heating, cooling and, depending on business needs, humidification and/or dehumidification. If there are 100 people in the office and if each person (according to guidelines) requires 8 l/s, then the ventilation requirement is 800 l/s.

If the ventilation fan is oversized, providing say 1,600 l/s, turning the airflow down to 800 l/s will give the maximum energy reduction for the business. If the fan is significantly oversized, a suitably sized fan meeting the building's needs will maximize energy reduction.

Using another example, heating 1 l/s of water by one degree consumes 4.2 kW. If the process or hot water boiler needs to raise the temperature by 10 °C – typical of many heating systems – it will consume 42 kW. If this occurs for 1 hour/day, the energy consumption is 42 kWh/day.

The knowledge of where and how much energy is used and what the business actually needs gives the maximum energy reduction potential. The business then finds a range of opportunities to close this gap.

Employees on the shop floor are able to identify more opportunities for improvement compared to managers at the top of the organization. In a study, shop floor employees and managers were given a short lecture about how to find opportunities for energy reduction. After the workshop, they were asked to identify the number of energy reduction opportunities contained in a fictitious case study. Then, they were asked to repeat the same exercise in their own workplace. In both cases, shop floor employees consistently identified twice as many opportunities as their managers. Do involve and include them at the outset.[1]

Turn off equipment when it is not needed. Turn down equipment by applying a variable speed drive (VSD) or similar tools. If there are several machines of different capacities, for example, boilers to provide heating, match the machine with the maximum efficiency at the required demand.

If replacements are necessary, gain a basic knowledge about energy systems to be able to communicate with consultants and contractors, and to confirm that the proposal meets organizational needs, i.e. it meets the demand without excessive spare capacity.

An example is General Electric's (GE's) 'Energy Treasure Hunt' – a hands-on employee engagement programme. It begins by putting together groups of cross-functional employees. These employees are trained to scrutinize energy use and identify inefficiencies.[18]

GE utilizes its own internal knowledge about the manufacturing process, its operational and maintenance expertise, and newly gained skills to map energy flows, assess energy use, track down wastes, identify opportunities for improvement and generate a list of projects and an action plan. If there are specific skills where in-house experts are not available, external expertise is sourced.

The idea behind the treasure hunt is about applying small and incremental improvements. It starts on a Sunday, observing and quantifying opportunities when the manufacturing plant is not

---

[1] For an introduction into employee engagement and behavioural change, please see Appendix A.

producing products. The treasure hunt continues on Monday and Tuesday when the manufacturing plant is ramping up to production speeds and into the production phase.

The energy treasure hunt teams own the opportunities. If necessary, formal cost–benefit analyses are carried out by the same team after the event. The same team is also responsible for implementing the opportunities. Even though there are no formal mandates for GE sites to carry out energy treasure hunts, more than 300 sites globally have carried out the hunt and have saved more than $150 million.[19]

Using the same building ventilation example described earlier, turning down the ventilation saves fan power. It also saves the energy associated with conditioning the air in terms of heating, cooling, humidification and/or dehumidification.

Many managers think that achieving large-scale energy savings will require large-scale capital projects. This is not true. Diageo North America – a group that generates $5.6 billion in revenue – is a prime example of defying this myth because it achieved 50 per cent energy savings using 'no-brainers' – no-cost and low-cost opportunities – and has delivered the savings 38 years earlier than planned.[20]

In 2008, Diageo North America began its initiative by setting up a team and carrying out the rigorous exercise of collecting and analysing ideas for improvement. These ideas were sorted by net CO_2 reduction and then by financial costs. The management was surprised by the number of no-brainers, ranging from lighting retrofits, boiler upgrades and installing VSDs to switching fuel oil to natural gas, and operating one boiler at full load instead of two at part load.

As another example, the Environment Agency, the UK's environmental regulator, had a target to reduce 33 per cent of its energy consumption across 40 of its buildings. First, the agency insulated its building fabrics, optimized hot water boiler operating times, and utilized high-efficiency lighting and voltage optimization. Secondly, it put in place an operating norm where no buildings are heated beyond 19 °C or cooled below 26 °C. The relatively simple solution gave an energy reduction of up to 34 per cent.[21]

When the organization becomes competent in attending to simple and low-cost opportunities such as those described above, the organization can build and develop its energy maturity by applying different types of improvement opportunities, as shown in Figure 1.1.

Chapter 1 Managing energy and auditing

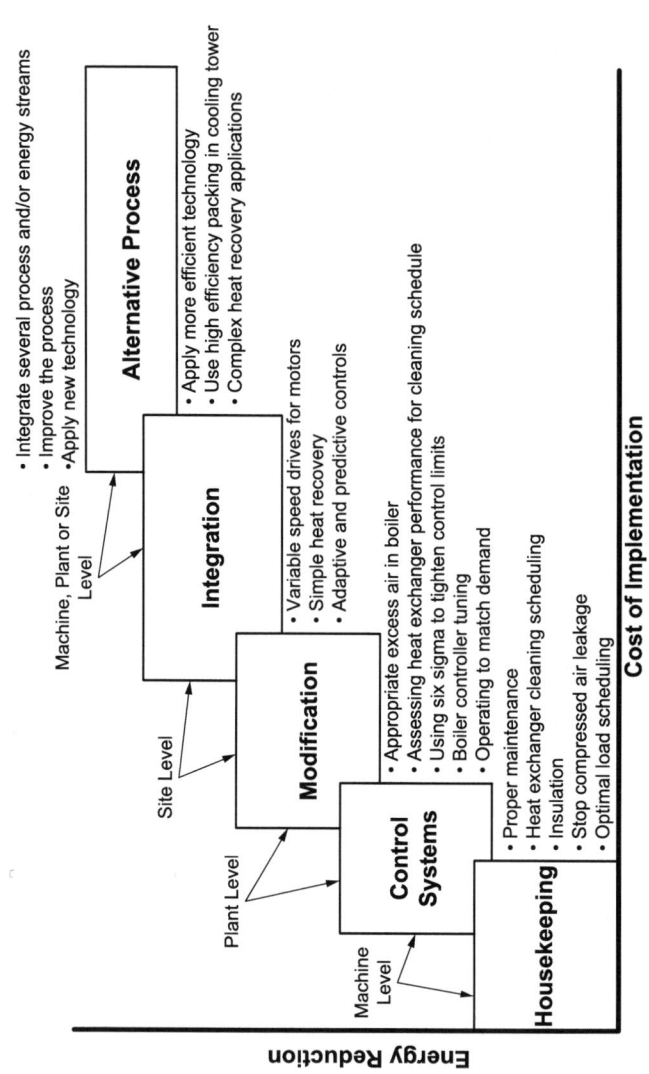

Figure 1.1: The energy maturity model
(Source: Oung, K (2013) Energy Management in Business: The Manager's Guide to Maximising and Sustaining Energy Reduction, Farnham: Gower)

The twin functions of energy management

Energy reduction and energy efficiency are not limited to the industrial and building sectors. Energy reduction in the transport services can also be achieved, especially when approximately 24 per cent of transport operations are operating virtually empty.[22] The Next Manufacturing Revolution lists 18 opportunities to reduce transport-based energy use.[23] Go-Ahead, the trains and buses company, and the joint efforts of Airbus, Singapore Airlines, Heathrow Airport and the UK national air traffic service (NATS) are some UK-based success stories in the transport sector.

Go-Ahead operates several transport franchises, such as London Midland Trains, Southeastern (trains), Oxford Bus Company, Plymouth Citybus, Brighton & Hove Bus and Coach Company, Go North East (bus company) and Gatwick Express. Go-Ahead carries more than one billion passengers each year. Working closely with Network Rail, Go-Ahead has deployed a regenerative breaking system on their trains which gives electricity savings of 8 per cent. The company also uses the telematics system to record the revving, idling, braking, acceleration and speed of every journey. Using extensive fuel-efficient driving, it has increased the fleet miles per gallon by 3.5 per cent with a further projection of 10 per cent coming from refinements to the driving methods.[24]

The Airbus A380 is the most fuel-efficient (17 per cent less fuel per passenger) and quietest (50 per cent less noise at take-off and up to 75 per cent less noise at landing) aircraft in commercial operation.[25] Heathrow Airport, on the other hand, is the busiest airport in the UK and the third busiest in the world.[26] Due to noise restrictions at Heathrow Airport, jumbo jets take off from the airport using the 'TOGA' procedure (maximum power) originally developed for the Boeing 747-400.[27] Airbus, Singapore Airlines, Heathrow Airport and NATS have analysed 10 to 12 months of operating data and flight path data to implement the fuel-efficient 'Flex' take-off procedure utilized at other airports. This procedure involves the A380s departing Heathrow using less power. Once the aircraft reaches a height of 1,500 feet, it uses flexible acceleration up to 4,000 feet, before continuing on its journey. This new procedure gives a fuel saving of 300 kg of fuel during every take-off.[28]

The topmost available form of energy maturity would be achieved from a step change in perspective by enlarging the scope of energy management into the organization's supply chain and/or by commissioning an alternative source of energy. Applying combined heat and power (CHP) plants and other low-carbon technologies falls into this category. Implementing them at this stage of energy maturity gives the lowest cost per ton of CO_2 and minimizes the initial amount of capital needed.

Depending on the industry cluster, the energy consumption by other companies in the organization's supply chain and, as a direct result, CO_2

Chapter 1   Managing energy and auditing

emissions, can be as high as 60 per cent. In the case of heavy industries, such as steel, aluminium, cement, ceramic, pulp and paper, this could be as much as 80 per cent.[29]

Walkers Crisps, a subsidiary of PepsiCo, is one of the largest snack food manufacturers in the UK. It discovered that farmers were storing potatoes in humid rooms in order to keep the potatoes' skin soft, to avoid them drying out, and thus meeting Walkers Crisps' purchasing specification. However, during the manufacturing process, the potatoes were dried and then fried. With a simple change in its potato specification, i.e. to purchase based on dry weight, plus other specifications, farmers are now able to reduce their energy and water costs and Walkers Crisps uses 10 per cent less energy to make its crisps.[30]

Some organizations roll out low-energy production methods or so-called 'green chemistry' for manufacturing the same product using less energy. CEMEX, a cement manufacturer, is a good example regarding green chemistry. More than 95 per cent of the CO_2 emissions from cement manufacturing come from the manufacture of clinker – a product from binding limestone, clay, iron oxide and alumina silicate at 1,450 °C. Since 2002, CEMEX has been adding a by-product from glass, steel and ash from a coal power plant. By doing so, CEMEX has been able to reduce the amount of clinker used by 30 per cent without affecting product quality. As a result, less energy is now required.[31]

If, when identifying and implementing energy reduction initiatives, internal employee resources are constrained, the use of external resources could be beneficial. Having said that, many experts, consultants and contractors tend to communicate using technically complex terminologies and, perhaps, terminologies that are alien to the organization.

An experiment was carried out three times using three separate groups of master's degree-level energy management students. They were given a fictitious energy management case study and asked to prepare an elevator pitch to the manager of the case study (represented by a MBA student). The results showed that the business degree student could not differentiate which pitch offered the best value for money.

From a layperson's perspective this makes little sense, as the energy experts are trying to sell to people who cannot fully understand them. Challenge the experts to use business English. Have their ideas documented so that their implementation can be checked. At the same time, apply business sense to minimize capital.

Finally, as with any operational management, managers need to ensure that all energy assets are maintained according to manufacturers' recommendations.

A simple illustration of energy efficiency economics

Let us consider the economics of a lighting energy reduction in a typical office building in central London. This building is upgrading all of its existing T12 fluorescent tube lighting. There are 100 light fittings; each fitting contains two 4 ft. fluorescent luminaries. The office is occupied 14 hours a day, 7 days a week. The lights are left on 24/7 and consume 8 kW of electricity. The lighting costs £6,475 per year and emit 32 tons per year of CO_2.

Assume that the building owner wants a totally off-the-grid solution and is able to work around land area limitations to implement LED lighting and wind turbines without first reducing energy consumption. The organization will need to install five vertical axis wind turbines (rated at 8.2 kW each) and a larger battery capacity to cater for periods of low winds. This saves £6,475 per year of electricity and costs the organization £5,613 per ton of CO_2 avoided.

If the organization was to consider energy reduction by simply implementing a policy of 'last one to leave turns off the lights', it would provide 58 per cent of the savings or £3,547 per year without any cost to the organization. Replacing the lighting with LED lighting gives an additional saving of £1,084 per year and costs the organization an additional £960 per ton CO_2 avoided.

Finally, going green at this stage by installing wind turbines would save an additional £1,466 per year with an additional cost of £4,335 per ton CO_2. In all, taking energy reduction into consideration gives the same savings but achieves this at a lower cost of £5,296 per ton CO_2 – £317 per ton CO_2 less by implementing energy reduction before going green!

Following on from the earlier case study of Diageo North America, it achieves a further 30 per cent energy reduction by using a different source of natural gas – landfill gas supplied by the utility company. Landfill gas is more costly but is a carbon-neutral source of fuel. Initial assessment indicates that an additional cost to purchase landfill gas could go up by $1 million per year and significantly eat into its profits.

A senior executive in Diageo – President of Diageo's Global Supply and Procurement and who sits on Diageo's Sustainability Council – gave the go-ahead and financial leeway to implement this. At a strategic level, it was the cheaper way for the group as a whole to achieve large-scale CO_2 reduction.[32]

As a further illustration, let us compare the economics of implementing CHP versus implementing CHP after energy reduction in an industrial setting. A fictitious manufacturing plant operates 12 hours a day, 5 days a

Chapter 1 Managing energy and auditing

week. It consumes 9.7 million kWh of natural gas and 18.9 million kWh of electricity over one year. This is equivalent to total CO_2 emissions of 12,125 ton per year.

If the plant was to implement CHP to augment heat and power needs before implementing energy reduction, it would install a 1.2 MW reciprocating engine. The project would cost £2.5 million to implement and save 1,087 ton CO_2.

The plant identifies nine energy reduction projects that would reduce the demands on natural gas by 8 per cent and on electricity by 21 per cent. At a cost of £0.2 million, the energy reduction projects would give a saving of 2,315 ton CO_2. Implementing a CHP scheme after energy reduction would result in a 0.7 MW reciprocating engine, cost £1.8 million and give an additional saving of 719 ton CO_2. By implementing CHP after implementing energy reduction, the plant would achieve an additional saving of 1,947 ton CO_2, using £0.5 million less capital.

Sustaining energy reduction

As mentioned at the beginning of this chapter, identifying and implementing opportunities to reduce energy consumption is only one part of the picture. In fact, 85 per cent to 90 per cent of organizations do not plan and implement their predetermined plans.[33] Without maintaining the achieved results, the project or initiative will become an ad hoc means to reduce energy consumption. In an ad hoc mode, managers implement projects to reduce energy reduction. When the energy projects are complete, managers refocus their efforts and attention on other pressing areas in the organization.

The newly installed and/or other equipment may be poorly installed, operated and/or maintained. There could be changes in customer requirements, changes in regulations, raw materials, etc. All of this could cause energy consumption to rise and managers would be unaware or unable to account for the increase in energy consumption. At this point, the cycle begins again.

According to Samantha Hodder, Go-Ahead's Group Communications Director,

> Go Ahead management made two things clear from the outset: first, that improving the organization's environmental impact would only work with collective action, and second that any strategy it developed would fail if it was not locally owned.[34]

The barriers to an organization sustaining energy reduction, manifests itself in:

- *lack of time and resources.* Everyone in the organization is focused on carrying out their individual tasks. Energy consumption is the second- or third-level priority. In terms of operating costs, raw materials and human resources are typically the first- and second-highest costs. Naturally, managers tend to gravitate towards these issues and to resolve the more pressing organizational matters;
- *operational losses.* Over time, new employees become accustomed to, and familiar with, the norms of operating requirements. This 'selective attention' means that, over a period of time, even a proactive, highly motivated and energy-efficient employee can be perceived to 'relax their standards' to fit in with other colleagues. This can make the difference between a well-run plant and building and one that has a lot of leaks, idle time, unreported issues and equipment failures;
- *maintaining the status quo.* Employees could also resist the adoption of energy efficiency and efficient ways of working because they are unfamiliar with the new ways of working. This is more prevalent when it involves non-technical personnel. For example, many food and health authorities require 'product-contact' water services used in the life sciences and food and beverage industries to be 'turbulent'. Even though a manager in a food and beverage organization may understand the concept of energy saving by utilizing a VSD, they may still resist it because they are not able to assess and quantify a sufficient level of 'turbulence' to comply with the 'turbulence' requirement set by the authorities. They may hide behind the fact by quoting the rule book, i.e. 'validation says it must be turbulent'. They may also resist taking action because of the bureaucracy involved, or volume of paperwork required, to make the change;
- *conflicting requirements.* Different stakeholders within an organization have different objectives and different interests. For example, production personnel are motivated to manufacture quality products, whereas maintenance personnel are motivated to take the time to carry out essential and qualitative maintenance. The production manager wants availability (or uptime) for manufacturing, whereas the maintenance manager needs time for good-quality maintenance. The continual conflict could mean that maintenance is never really completed satisfactorily or that allocated time for maintenance is a moving feast;
- *uncertainty in user requirements.* The actual capacity and capability of all the installed equipment is normally larger than that specified in the original specification. This could be due to an unknown or vague user requirement. For example, a building designed for a fixed number of occupants will probably have a smaller ventilation system compared to a similar building with an unspecified number of occupants. It could also be because, as the approved design is passed from designer to manager to purchasing department to supplier to manufacturer, each adds a 'safety' margin. For example, a pump, if

Chapter 1   Managing energy and auditing

unchecked, is typically 50 per cent to 100 per cent larger than that specified. The consequence is that the pump is more costly to purchase and will operate at a lower power factor. The worst-case scenario is that the organization may be fined for having a lower power factor, or it may have to purchase a power quality correction machine to compensate for the lower power factor;
- *implementation shortcomings*. Due to limited resources and the urgency of implementation, many energy reduction initiatives are started prematurely. This situation could arise from:
  o implementing a design that has not been fully assessed and designed;
  o approving funds for a capital-intensive project based on estimates;
  o not fully appraising the business risks, with an impact on other parts of the business; and/or
  o retrofitting part of a machine without due consideration to operability and maintainability of the whole machine and/or process;
- *stuck in a perpetual cycle for perfectionism*. Sometimes initiatives can become stuck in a cycle where managers are too focused on getting every detail perfect. This need for perfectionism may cause the initiative to be delayed or stalled, or a lot of time and resources to be spent that far exceeds the economic benefit. Furthermore, there is no guarantee that the initiative will be perfect from the start and will not need tweaking or fine-tuning.

Overcoming the barriers to maintaining and sustaining a well-implemented energy reduction project relies on the integrated systems and processes a company uses to govern its day-to-day operations – a management system. A management system is an 'integrated set of processes and tools that a company uses to develop its strategy, translate it into operational actions, and monitor and improve the effectiveness of both'.[35]

The role of top management

To implement and generate significant value from implementing a management system requires senior management to demonstrate that saving energy is not another initiative but a new way of working and that it is part and parcel of organizational performance. A committed and involved senior management allocates and authorizes the relevant resources, disarms the barriers to energy efficiency and ensures that the management system works.

Diversey, an industrial and commercial cleaning and sanitization company, now part of Sealed Air, has been implementing a variety of ad hoc energy reduction projects. In 2008, Diversey pledged to reduce its energy consumption by 8 per cent below that of 2008 (as a base year) by 2013. Initially, the team at Diversey identified 120 improvement opportunities, of which 30 met the standard hurdle rate, to achieve a total energy reduction of 8 per cent.

Senior management at Diversey began to examine these 120 opportunities and found that, due to a mismatch of priorities, resources and incentives, their managers lacked the capabilities and motivation to maximize energy reduction and select the minimum resources required to meet the corporate target. The active involvement of senior management in managing energy, in assessing projects, realigning priorities and incentives, and ring-fencing corporate finance allowed Diversey to implement 90 opportunities, achieving a 25 per cent energy reduction with $5 million less capital.[36]

The example above shows the benefit of having senior management commitment and involvement in driving energy savings. A committed senior management also steers its perception on long-term sustainability, how it chooses to manage energy consumption, how it measures and reports energy at boardroom levels, and its strategic plans on climate change adaptability. Senior managers who do not incorporate sustainability face the risk of boardroom and/or shareholder revolt, demanding action on sustainability.

ISO 50001, *Energy management systems* — *Requirements with guidance for use*

ISO 50001, *Energy management systems* — *Requirements with guidance for use*, published by the International Organization for Standardization (ISO) in late 2011, is a management systems standard for managing energy consumption. It contains all of the features needed to identify and prioritize opportunities to reduce energy consumption. It also contains the features necessary for sustaining reduced energy consumption.

An ISO 50001-based energy management system is a 'set of interrelated or interacting elements to establish an energy policy and energy objectives, and processes and procedures to achieve those objectives' (the continual improvement of energy performance).[37] The definition of an ISO 50001 energy management system parallels the definition of a management system: a set of integrated processes and systems to deliver a desired outcome.

Organizations need to understand the fundamental principles of the various management systems and recognize that they are the same but with different 'focuses'. For ISO 50001, the focus is on energy. Figure 1.2

shows the activity-based features of ISO 50001. In terms of sustaining energy reductions achieved, this means:

- planning operations and maintenance in such a way as to minimize energy consumption;
- carrying out planned operations and maintenance according to plan;
- measuring and checking that projects are delivering the stated benefits;
- measuring and checking that the planned activities are effective in controlling energy consumption;
- continually monitoring and identifying new opportunities to minimize energy consumption.

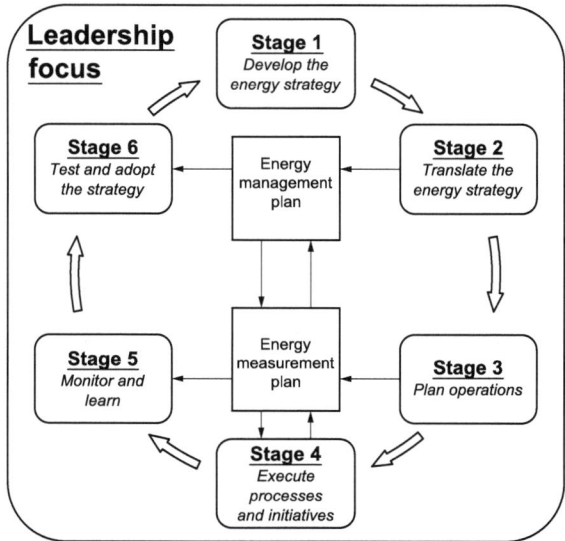

Figure 1.2: External features of ISO 50001

An ISO 50001-based energy management system is also a management system where an organization can seek third-party assurance to prove that its practices conform to an energy management systems standard, and that it is creating value from its implementation. The robust verification processes used by the independent body remove information asymmetry between organizations in terms of their energy management efforts and signify that their management practices contain the necessary processes to ensure continued and longer-term energy reduction.

In a little over three years since the publication of ISO 50001, organizations in the EU are leading the way in implementing and

ISO 50001, Energy management systems

adopting the standard as a holistic way to manage energy. Similar to ISO 9001 and ISO 14001, the number of companies certified to ISO 50001 is expected to rise, either because of the need to compete with global organizations or because regulations would specify ISO 50001 as a precursor for compliance. (See Figure 1.3.)

When implementing a management system, many think of it as an independent project, following a rigid process to meet the requirements word for word. For example, an organization certified to ISO 9001, ISO 14001, OHSAS 18001 and ISO 50001 can end up spending time and resources operating five independent management systems (the fifth being the way the organization operates on a daily basis).

Inevitably, many organizational activities are focused on resolving nonconformities and collecting evidence to meet the requirements before the next management system audit. Therefore, organizations become caught in a spiralling loop of constant catch-up from audit to audit. Many are suffering from initiative fatigue and are not able to apportion successes or otherwise to specific initiatives.

Regardless of which management systems the company uses – the ISO 50001-based energy management system, other 'plan–do–check–act' (PDCA)-based management systems, Robert Kaplan–David Norton's Balanced Scorecard system,[38] John Kotter's change management [39] framework, Scott Keller and Colin Price's *Beyond Performance* [40] model or many other management systems models – there are several inbuilt features to encourage sustainability of energy reduction. A good and thorough understanding of these principles allows organizations to distil the essence of governance and operate one management system. This is shown in Figure 1.4 and is listed below.

- Set clear and high aspirations for change – poorly defined and 'grey' areas create a space for misinterpretation and drifts away from the original intent. Setting a clear vision for improvement, along with achievable results, creates a pathway for employees to carry out their work. This is highly motivating.
- Simplify the language and terminology – using complicated and bombastic words makes the speaker sound important but leads to nothing because people do not relate the terminology to their work. Using simple and common terminologies within the organization allows employees to assimilate the message with little misunderstanding. Employees are, therefore, able to work towards a common goal, in a consistent way.
- Communicate authentically through words and through actions – being inauthentic is very easy to identify and significantly undoes the trust and respect of employees. Leaders and managers leading by

Chapter 1 Managing energy and auditing

Figure 1.3: Growth in demand of ISO management systems standards

(Source: Data adapted from Sector Forum for Energy Management)

ISO 50001, Energy management systems

'visible and audible' example sends a powerful message to all employees that the management sees value in energy reduction efforts, and is prepared to do the tasks and achieve its goals.
- Create opportunities for colleagues and employees to get involved – employees enjoy quick wins and successes. Create opportunities for contribution and congratulate them on jobs well done. It breeds happiness and motivates employees to do more.[2]
- Provide continuity of management communication and focus – implementing many and disparate management change programmes slows down the change efforts. Consistency and continuity renew and rejuvenate energy reduction efforts by creating the assurance that the organization is not losing sight of, and continues to lower, energy consumption. They also help to break down a silo mentality and encourage the organization to think and act towards the overall organizational vision, rather than towards departmental visions.

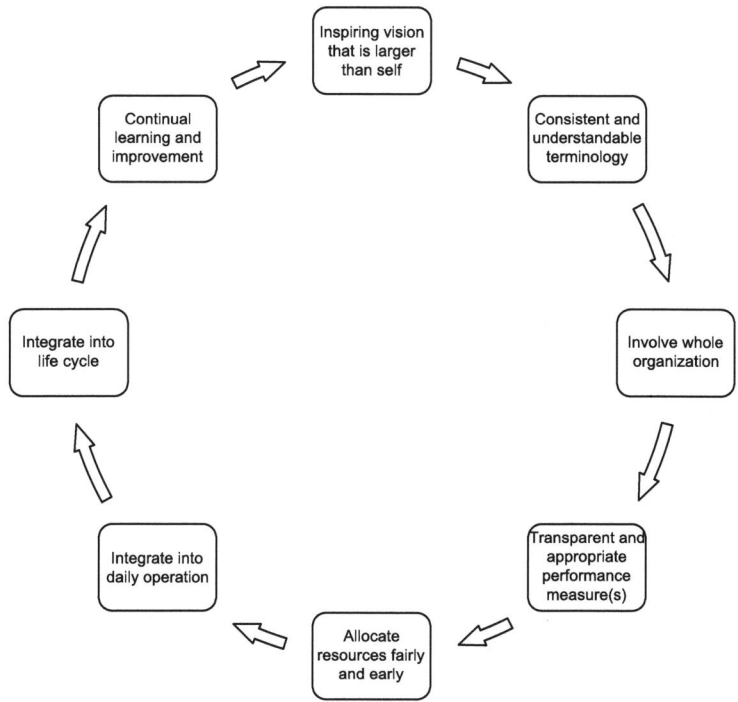

Figure 1.4: Intrinsic features of ISO 50001

---

[2] For an introduction into employee engagement and behavioural change, please see Appendix A.

Energy Audits 21

Organizations that truly understand these intrinsic and external features of a management system will find that more than 50 per cent of the processes are common among all of the management systems and are also common in current business practices.

The uses of energy audit in energy management

In general, many organizations are aware of the opportunities available to save energy in their workplace. An energy audit is a systematic and methodological approach to identify and quantify the benefit of energy savings. In companies that have an ISO 50001-based energy management system, there are several areas where an energy audit can complement the organization's efforts in managing energy. One of the more obvious links to energy auditing is the input into an energy review to:

- identify the energy sources used by the organization;
- evaluate the quantum of energy consumed;
- identify significant energy users[3] (equipment, systems or processes etc.) and the people working with these significant energy users;
- identify the variables that cause the energy consumed by the significant energy users to vary;
- determine the appropriate energy performance indicator and the energy baseline based on the significant energy users;
- find opportunities to improve the energy performance of the significant energy users.

An energy audit can also be used within an ISO 50001-based energy management system to assess and update the energy review following major changes and modifications within the organization.

An organization may also choose to carry out energy audits on specific equipment or groups of equipment (e.g. pumps, fans, air compressors, boilers or process use), using specialist consultants. The organization then pulls together all of the various audit findings and completes the requirements of the energy review on its own.

Other reasons why an organization will carry out an energy audit include to:

- comply with regulations, such as the ESOS or EED (see Chapter 4);
- apply a disciplined approach to quantifying the cost and benefit of a specific opportunity to improve energy performance with a view to capital sanctioning;

---

[3] ISO 50001 defines significant energy use as an item, equipment or machine, or group of equipment that consumes a substantial quantity of energy and/or offering considerable potential for energy improvement.

The uses of energy audit in energy management

- identify and quantify energy reduction opportunities in order to plan for the following year's capital budget;
- identify and incorporate energy efficiency into planned refurbishments, modifications or end-of-life asset replacement;
- carry out independent due diligence on the work and services provided by an energy performance contracting (EPC) company or an Energy Service Company ESCO;
- carry out an independent assessment of an energy reduction proposal requiring high capital expenditure in order to minimize investment and business risk;
- identify opportunities for improvement as part of a wider, organizational sustainability and/or environmental assessment, e.g. as part of an assessment including energy, water, waste and idle time;
- identify opportunities for improvement as part of a wider supply chain or value chain assessment.

Lastly, an energy audit can also form a site-wide or property-wide master plan for future strategic development, incorporating energy efficiency, environmental and sustainability parameters into the organization's plan. A good example is Pfizer's energy master plan for its Freiburg site. Pfizer, a US pharmaceutical company with manufacturing and research and development (R&D) facilities across five continents, has an ambitious energy and climate change programme. The company, having achieved a 43 per cent reduction in energy consumption relative to its sales in 2007 from a baseline of 2000, announced that it was going to reduce its energy consumption by a further 20 per cent by 2012.[41]

The site in Freiburg, Germany, achieved its reduction by designing and fastidiously following an energy and resource conservation master plan. The plan identified and assessed a portfolio of about 200 projects according to their cash flow implications for the site, in terms of energy, engineering, maintenance, profit, risks and the company's future investment plans.

The project ranged from no-brainers with low, upfront investment and relative low risk, such as insulation, turning off and turning down air conditioning systems, heat recovery, adiabatic cooling and automatic power shutdown procedures to high-end building renovations to improve building fabrics and utilizing renewable energy sources such as CHP plants, biomass boilers and photovoltaic systems.

Energy Audits

Chapter 2   Energy audit standards

Energy use in buildings, processes and transport accounts for nearly 100 per cent of all energy consumed. Figure 2.1 shows the energy consumption breakdown in the UK, the EU and the USA in 2010. As described in Chapter 1, organizations wanting to reduce their energy consumption first need to identify opportunities to do so. An energy audit is one of the most commonly used tools to identify opportunities for reducing energy consumption and to enrol the organization to take action.

In the UK's and EU's buildings sector, new buildings represent only 1 per cent of the total available stock.[42] Therefore, the opportunities to save energy in buildings are limited to retrofitting existing buildings. Up to 60 per cent of energy consumption in a building is used for heating, cooling, ventilation and air conditioning, where energy savings of up to 30 per cent can commonly be found. The rest of a building's energy savings comes from improving the building fabric and from understanding and minimizing how occupants use energy. As such, energy auditing the building sector follows a similar thought process and is well established.

The manufacturing and industrial processes use many different pieces of equipment, unit operations and systems. They also have diverse operating conditions, each specific to a product or a group of products. They are also very prone to energy price volatility, customer demands and product innovations. The breakdowns of energy use and energy consumption in the manufacturing and industrial sectors are, therefore, less straightforward. Having said this, as shown in Figure 2.2, an annual energy reduction of up to 30 per cent can be found in many manufacturing and industrial sector organizations.

Carrying out an energy audit in the transport sector is less common and patchy. This is because: (1) there are significantly fewer organizations in transport compared to building, manufacturing and industry, and (2) many of the energy audits carried out by transport companies are around the buildings they occupy. Chapter 1 has highlighted examples of energy reduction opportunities for transport operations. Another good example comes from Matthews Coach Hire in Ireland.

Chapter 2   Energy audit standards

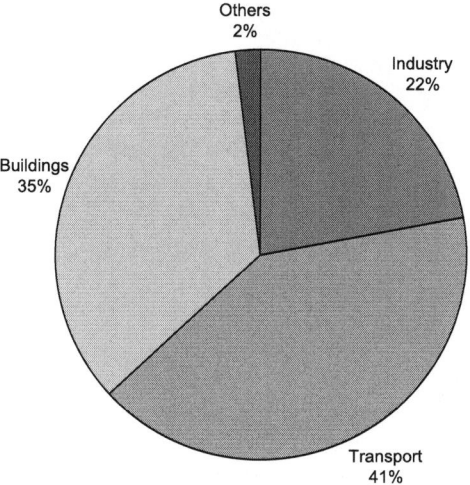

Figure 2.1a

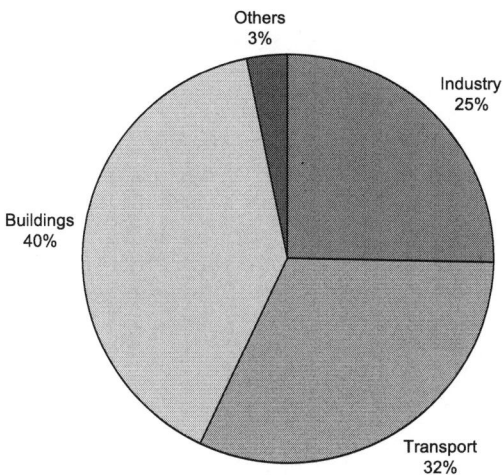

Figure 2.1b

Chapter 2   Energy audit standards

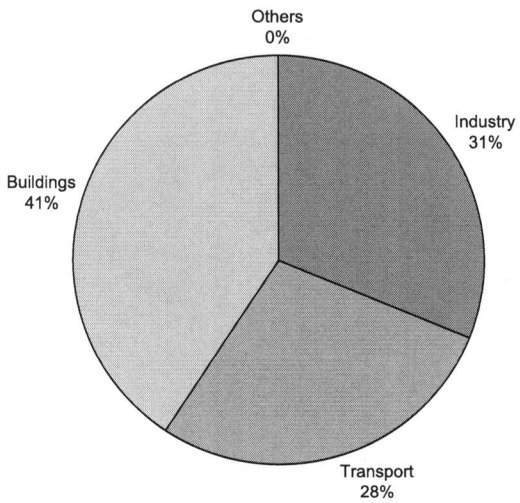

Figure 2.1c

Figure 2.1: Final energy consumption by sector (a) in the UK, (b) in the EU and (c) in the USA in 2010

(Source: Eurostat, European Commission (2012) Eurostat Pocketbooks: Energy, transport and environment indicators, Luxembourg: Publications office of the European Union and the US Energy Information Administration (2013) Monthly Energy Review October 2013, Washington: US Energy Information Administration)

Matthews Coach Hire has a fleet of 34 coaches providing coach tours in Ireland. Matthews first carried out a transport-based energy audit in 2007 and received its ISO 50001-based energy management system certification in 2012. Throughout this period, it has, to a large extent, operated the same fleet, operating the same route, and using the same drivers. Over the five years since the energy audit, its energy performance indicator has dropped year on year from 35 l/100 km in 2007 to 20 l/100 km in 2013 – a 43 per cent improvement![43]

Standardizing energy audits

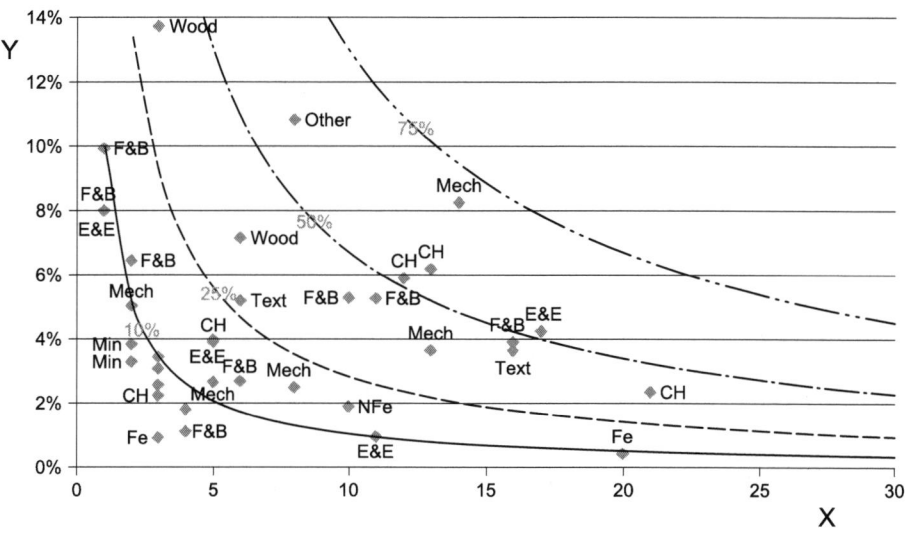

Figure 2.2: Energy reduction potential across several manufacturing sectors

(Source: Lavery, G, Pennell, N, Evans, S and Brown, S (2013) The Next Manufacturing Revolution: Non-Labour Resource Productivity and its Potential for UK Manufacturing, July 2013, Lavery Pennell, 2degrees and The Institute for Manufacturing)

Standardizing energy audits

The word 'audit' itself is a very loosely defined term in energy management. An 'audit' could be known as 'analysis', 'diagnosis',

Chapter 2   Energy audit standards

'review', 'assessment', 'survey', 'walk-through survey', 'opportunity finding', etc. Some ESCOs refer to them as 'opportunity days', 'ESCO proposal development', etc.

When these energy audit models are analysed, there are many variations, unique features, approaches, work methods and scopes of work specific to the service provider. In some countries, legislation or regulations specify a requirement for, and applicability of, different types of energy audits. Some countries have a single standard or guideline. Others may have several guidelines with slightly different scopes and specifications.

Some service providers and organizations use the term 'audit' to mean a cocktail of tools and techniques with varying levels of applicability, detail, thoroughness and degree of confidence in the findings. This range could include:

- an energy management system assessment – analysing the effectiveness of an organization's governance of energy consumption;
- energy accounting – the checks and balances of energy an organization purchases and its consumption;
- energy monitoring and targeting – the comparison of an organization's energy consumption with its activities, i.e. production volume, trade, etc.;
- energy benchmarking – comparing the energy consumption or energy efficiency of the organization or specific equipment with a similar installation or better;
- energy opportunity finding – finding opportunities for reducing energy consumption within the organization;
- energy data collection and reporting software – software to collect data for later analysis;
- energy audit software – preprogrammed software to analyse inputs by a user that recommends a predetermined opportunity for improvement.

As each service provider uses the same word, 'audit', to mean a different combination of things, the level of competence and knowledge in carrying out an 'energy audit' also varies. On the one hand, there are simple energy audits based on a checklist, a tick-box exercise or filling in preprogrammed spreadsheets or software. These tend to be low cost and the energy auditors are less skilled in the operations of the equipment and machinery. On the other hand, there are very skilled specialists: people who are very knowledgeable regarding specific processes, e.g. compressed air, boilers and solvent recovery units. These energy audits tend to be limited to their specific competence and more costly.

Because of these variations, it is very difficult for organizations to identify the best proposition that will meet the organization's needs and that is

Standardizing energy audits

value for money. When comparing different proposals from different professional service providers, some of the issues managers face are as follows.

- What is the proposed scope of work? How in-depth is the proposed energy audit?
- What is the process of carrying out the proposed energy audit?
- What involvement is required from the organization?
- Are the audit findings based on rules of thumb, assumptions, benchmarked data or measurements?
- Are the information and savings projections provided trustworthy?
- Will there be any risks and interruptions to existing organizational requirements?
- Are the opportunities real and achievable for the business? Are there any safety margins or fudge factors built in?
- How will the organization know that any proposed improvement opportunities are what the organization really needs and are not a scam to sell a product?
- Will the proposed improvement opportunities be credible? Are they going to affect the operation of other parts within the organization?
- Are there any other non-energy-related benefits from the proposed improvement projects?
- Is the service provider competent in its work and knowledgeable about the plant or machinery and its interdependencies with other equipment?

This is where a Europe-wide and globally accepted standard on energy auditing comes in handy. It allows the common elements of an energy audit to be standardized in a document. Within the standard, work specifications, the process of carrying out an energy audit, the roles and responsibilities of energy auditors and organizations, and the output from an energy audit can be documented.

A harmonized and globally relevant standard on energy auditing increases the visibility of quantified opportunities for energy reduction, and facilitates organizations to understand the outputs from an energy audit and use the findings to save energy. From an economic perspective, standards also allow fair competition and provide a basis to compare proposals from service providers in a like-for-like manner. Should the organization wish to, the process of carrying out an energy audit can also be audited and verified.

Ingersoll Rand is a $12 billion global corporation with brands such as Club Car, Hussman, Ingersoll Rand, Schlage,[4] Thermo King and Trane. With 100 manufacturing sites globally, it uses three-tier energy audits (treasure hunts, expanded energy audits and systems-specific audits) across its sites to identify opportunities for improvement by applying a

---

[4] Hussman and Schlage are no longer part of Ingersoll Rand.

sound discipline of generating business cases with the right level of detail for each of the opportunities. This has helped the company to overcome three of its biggest barriers:

1) creating visibility in energy consumption among employees and management;
2) providing adequate information tied to each type of improvement; and
3) engaging management buy-in for resources to minimize energy consumption.

The energy auditing events also serve as a training opportunity for employees and managers to learn how the organization uses energy. The programme has saved Ingersoll Rand more than $5 million since its inception in 2005.[44]

As another example, Marks and Spencer (M&S) is one of the UK's largest retailers with sales of over £9 million. In January 2007, M&S rolled out 'Plan A' – a 100-point plan to improve a range of issues, such as health, ethical business practices and climate change, over a period of five years on a cost neutral basis. This included:

- reducing energy consumption from stores by 25 per cent on a per square foot basis;
- improving the energy efficiency of offices and warehouses by 20 per cent;
- 20 per cent of new buildings to include on-site energy generated from renewable sources;
- all of its buildings' electricity consumption to be from renewable sources;
- putting in place a travel policy that favours rail travel instead of flights;
- using company cars that are of diesel and hybrid variants;
- using trucks that are 10 per cent more energy efficient;
- converting existing trucks to the more energy-efficient Euro IV and V engines.

M&S's requirement for cost neutrality instilled discipline in all M&S managers to appraise all energy efficiency investments, and focused their minds on finding and implementing tried and tested opportunities for improvement. In addition to the reduced energy operating costs, Plan A also allowed M&S to create brand and reputational benefits.[45]

European series of standards on energy auditing (EN 16247 series)

A pan-EU team of energy audit experts were assembled on 11 September 2009 to develop a standardized energy audit method in Europe. The

Standardizing energy audits

development was officially endorsed by the European Commission (Mandate M/479 EN) on 13 December 2010 as an instrument to [46]:

- reduce uncertainties in the expectations of an energy audit;
- define the terminologies used in an energy audit;
- reduce the risk of energy efficiency investments to organizations;
- ensure confidence in the output from an energy audit;
- allow fair competition of energy auditors across the EU.

The mandate also expanded the scope of work to be completed by the European energy audit experts by specifying a four-part standard to cover:

1) general;
2) buildings;
3) processes; and
4) transport.

Part 1: General requirements

As a first task, the experts surveyed the energy audit practices in the EU to develop a clear and concise picture of energy auditing practices in Europe. The survey gave an insight into the common features and processes in an energy audit, to help structure the energy audit standard. EN 16247-1 specifies the terminology, the process of an energy audit, the task specifications at different stages of an energy audit, and the deliverables from an energy audit that are common to all parts of the series.

In line with the EU's energy efficiency aspirations, the standard is developed with energy efficiency in mind.

Part 2: Buildings

The Energy Performance of Buildings Directive (Directive 2002/91/EC and recast in Directive 2010/31/EU) has been in force since 2002 and there is a wealth of standards supporting the directive. EN 16247-2 specifies the additional requirements for carrying out an energy audit of buildings' structure, fabric, heating and cooling, ventilation and air conditioning, hot water, light and lighting, associated controls and other building services.

Part 3: Processes

EN 16247-3 specifies the additional requirements for carrying out an energy audit of industrial processes, systems and equipment. As there are

Energy Audits

many variations in equipment and operating conditions, this standard has a wide range of applicability and references the ISO 50001 energy management systems standard.

Part 4: Transport

EN 16247-4 specifies the additional requirements for carrying out an energy audit for a range of transport services, from heavy-duty vehicles to public transport systems and logistics.

Part 5: Competency

Towards the end of 2012, although not mandated to do so, the EU team of energy audit experts decided to create EN 16247-5 to generalize the competencies of carrying out an energy audit as specified in EN 16247-1. The UK's interpretation of EN 16247-5 is PAS 51215.

International standard on energy auditing (ISO 50002)

The problems and issues experienced by organizations when commissioning, and in carrying out, an energy audit are not limited to the UK and the EU. Globally, it is also relevant. An international team of energy audit experts had its first meeting on 31 November 2011 and utilized EN 16247-1 as a starting point.

ISO 50002 sets out the minimum requirements for conducting an energy audit to identify opportunities for cost-effective improvement in energy. Instead of purely looking for energy efficiency opportunities, as in the case of EN 16247, ISO 50002 aligns with energy performance, including energy use, energy consumption and energy efficiency, in a similar fashion to ISO 50001.

> This International Standard specifies the process requirement for carrying out an energy audit in relation to *energy performance*. It is applicable to all types of establishments and organizations, and all forms of energy and energy use.
>
> This International Standard *specifies the principles* of carrying out energy audits, *requirements for the common processes* during energy audits, and *deliverables for energy audits*....
>
> ISO 50002, Clause 1 (the emphasis is the author's own)[47]

By bringing the standards into an international arena, ISO 50002 incorporates the good practices from other non-European countries and makes it a valuable tool for organizations wanting to carry out an energy audit as part of their energy management efforts.

Different uses of energy audit standards

Although ISO 50002 standardizes the common processes and output from an energy audit, it recognizes that there can be differences in terms of scope, boundaries and audit objective when carrying out an energy audit. This is necessary to meet the varying needs of organizations when carrying out an energy audit. This is especially true for small organizations, where following EN 16247's elaborate and detailed approach would be very costly and discourages them from carrying out an energy audit.

ISO 50002 identifies three types of energy audit that can meet the requirements of the standard. Type 1 is the basic, or entry-level, energy audit with type 2 and type 3 increasing in depth, thoroughness, time and cost:

- *type 1* – basic energy audit to identify high-level opportunities and to quantify low-cost and low-level risk opportunities. It may also identify opportunities to be developed at a later stage (type 2 or type 3);
- *type 2* – energy audit to firm up details and implement medium-cost or medium-risk opportunities. It may also identify opportunities to be developed in a type 3 energy audit;
- *type 3* – energy audit to quantify the feasibility and viability of high-risk, high-cost opportunities.

Table 2.1 shows a side-by-side comparison of the three types of energy audits.[48] All three types of energy audit can meet the requirements of the energy audit standard. An organization may decide on any single type of audit, start with a type 1 audit and use the results to progress to one of the other types, or 'mix and match' the details necessary to meet the needs and objectives of the organization. Table A.1 in ISO 50002 gives an in-depth description of the three types of energy audit.

Chapter 2   Energy audit standards

Table 2.1: Different types of energy audit and their application

	Type 1	Type 2	Type 3
In-company energy reduction	Audit of key pieces of equipment, e.g. motors, pumps and/or steam traps. Suitable for organizations with a small number of energy users. High-level snapshot for medium- and large-sized organizations to identify and prioritize efforts. Savings are based on equipment improvement only. Carried out by equipment supplier or by energy auditor as a precursor to type 2 work. Utilizes overall energy data, 'rules of thumb', 'benchmarks' and 'checklists' in estimating savings and capital costs. Energy savings and capital costs are typically to an accuracy of ±50 per cent. Low-risk and low-cost opportunities are generally implemented at this stage. Typically a half to one day's site work.	Audit based on a complete system of energy use, e.g. manufacturing line, hot water system and/or ventilation system. Assesses the interaction between several systems of energy use and energy or utility supply, where savings are based on optimizing the energy consumption of the whole system rather than individual components. May also be commissioned as an independent due diligence of existing energy service providers and/or to review proposals for energy savings from other energy sources. Carried out by competent personnel with knowledge and experience of the system. Energy savings and capital costs are typically to an accuracy of ±25 per cent. Some opportunities are implemented at this stage; others may be progressed to type 3. Typically two to five days' site work.	Audit based on a whole building and/or facility to assess the interaction of all systems and energy use, and to identify opportunities to reduce overall energy consumption. Typically used for a detailed audit of a specific energy reduction opportunity of a detailed design, involving high business or capital risk. May also be used by EPC companies prior to financing and guaranteeing energy reduction projects. Audits are carried out by specialists with in-depth knowledge and experience of the system or opportunity to be audited. Savings are based on detailed measurements and calculations. Energy savings and capital costs are typically to an accuracy of ±10 per cent. Depending on the size, can range from one month to one year.

Energy Audits

Empire State Building

One of the best-known examples of an energy audit is the iconic Empire State Building, New York. Opened in 1931, the Empire State Building draws in almost 4 million visitors every year to its observation tower. The building is also home to 2.8 million square feet of leasable space. At its prime, it was an exemplar of the mechanical age, with the tallest elevators and broadcasting masts atop the building. In 2006, after 75 years in continuous use, the Malkin family and the Leona Helmsley Estate (both are the owners of the building) decided to embark on a $500 million top to bottom renovation, upgrade the aging infrastructure and make it an exemplar energy-efficient building.

The motivation to include energy reduction was to:

1) prove the cost-effectiveness of energy efficiency retrofits;
2) reduce GHG emissions;
3) address other aspects of sustainability, namely water, recycling and reuse of building materials; and
4) examine the economics of capital improvement based on cash flow from reduced energy costs.[49]

In 2008, Anthony Malkin assembled a team of experts consisting of Clinton Climate Initiative as the facilitator and convenor of the project, Johnson Controls, Inc. (JCI) as the performance contracting company, Jones Lang LaSalle as the programme manager and Malkin's representative, and Rocky Mountain Institute as a not-for-profit, independent peer review. By this time, major capital projects were under way to refurbish the building.

Four criteria were discussed and agreed between the four parties:

1) if the energy savings did not pay back the capital investment, JCI would pay the difference;
2) all of the team were to forgo the payment for developing the energy reduction programme of work, opting to be paid in other works;
3) anyone should be able to copy the energy reduction process of the Empire State Building and reap similar benefits;
4) the retrofit had to emphasize energy reduction and energy efficiency in order to keep the cash flow and jobs in the USA.[50]

The Empire State Building had an electrical maximum demand of 9.6 MW and emitted an equivalent of 25,000 ton CO_2 every year. This equated to an energy cost of $11 million per year ($4/sq. ft.) and an energy performance indicator of 3.8 W/sq. ft. – the average within Manhattan Island. The energy audit took seven months to complete and was executed in four distinct phases:

Chapter 2   Energy audit standards

- *Phase 1: Identify opportunities.* This phase took almost one and a half months to complete while the team surveyed the building's mechanical and electrical systems, calculated energy usage and developed a theoretical minimum energy consumption based on the existing tenants. The team also carried out a gap analysis using Leadership in Energy & Environmental Design (LEED) and Green Globes® criteria. More than 60 opportunities were identified. The team also took an active part in modifying the existing capital project strategies. Of the 23 capital projects, 4 were put on hold, 6 were modified to achieve higher energy reduction, and 6 benefited from a capital cost reduction. The findings at this stage indicated an energy reduction of between 15 and 25 per cent, with a payback (including capital cost savings) of 5 years.
- *Phase 2: Evaluate opportunities.* This phase also took one and a half months, while the team took detailed measurements in order to create an insight into how the building's tenants used energy, and its pattern, and refined the energy model developed during Phase 1. Conventional thinking regarding heating, cooling and ventilation strategies, interior wall designs and lighting strategies were challenged. This phase also saw the introduction of life cycle costing and tenant agreements to reduce energy consumption. The quantified opportunities for energy reduction increased to between 40 and 50 per cent.
- *Phase 3: Create packages.* During this phase, the team focused on tenant engagement by:
    1) installing tenant sub-metering;
    2) working with tenant energy champions to reduce their energy consumption;
    3) providing tenants with online energy reduction education material; and
    4) providing real-time energy consumption feedback.
- *Phase 4: Model iteration.* By utilizing portfolio analysis and sensitivity analysis, the team's primary task was to minimize capital expenditure in order to maximize returns on investment from all available opportunities. Reasonable cost escalation factors, such as fuel, construction, inflation, discount rates and rent premiums were incorporated into the 15-year life cycle costing. The original 60-plus opportunities were narrowed down to a portfolio of 17 projects, reducing the maximum demand by 3.5 MW and giving a 37 per cent reduction in energy, covering:
    o using digital demand control (9 per cent);
    o redesigning the layout to maximize tenant exposure to daylight (6 per cent);
    o removing electrically powered ventilation fans and using mechanical dampers to enable natural ventilation [51] (5 per cent);
    o retrofitting chillers (5 per cent);

Different uses of energy audit standards

- o installing a third glazing for the windows during refurbishment (5 per cent);
- o the energy management efforts of tenants (3 per cent)
- o installing radiative barriers (2 per cent); and
- o occupant-led ventilation (2 per cent).

Two of the portfolio of projects started at the end of Phase 1. Five separate energy performance contracts were agreed between JCI and the Empire State Building Company. These were to deliver a total of 61 per cent of the savings and have a capital cost of $20 million. In addition, the work to improve the energy performance also led to an improvement in the bottom line. The average rent in 2006 was $26.50/sq. ft. The Empire State Building Company began signing new leases averaging between $40/sq. ft. and $60/sq. ft.

Other approaches

Mixture of in-house and external energy auditors

An organization may also prefer to carry out an energy audit using a combination of internal and external resources. Using a similar approach to GE's and Ingersoll Rand's treasure hunt, GlaxoSmithKline (GSK), a pharmaceutical company based in the UK, utilizes a home-grown methodology called 'Energy Kaizen' to identify energy reduction opportunities. Prior to the kaizen event, GSK utilizes teams of energy consultants to carry out energy audits of key energy use that are less familiar to GSK site employees. Their function is to collect energy use data, carry out preliminary data analysis of energy consumption, and assess the efficiency and opportunities of specific, non-production-based energy use, such as boilers, electricity, ventilation systems, chillers, steam traps, and motors and driven systems.

Using a week-long, facilitated, kaizen workshop approach, a cross-functional team – consisting of cross-functional GSK employees, and the team of external experts – reviews the energy use pattern, understands the operating patterns, and operational and maintenance controls, and carries out site walks during the daytime and at night-time, to identify and quantify opportunities for improvement. All opportunities for improvement are prioritized into an implementation plan.

At the end of the week, the plan is presented to the site's management. The site's management decides and plans for additional work to develop the details of the plan and to approve it for implementation. Since their inception in 2009, Energy Kaizens have helped GSK sites like Irvine, UK to reduce their CO_2 emissions due to energy consumption (in Irvine's case, by 24 per cent). A similar programme for water has been developed and deployed.[52]

Resource efficiency

Increasingly, organizations are realizing that saving energy can also lead to a reduction in the need for other organizational resources. As shown in Figure 2.3, organizations are looking beyond a pure energy reduction initiative to one that is a combination of energy, water, waste and other sustainability parameters.[53] This is collectively known as sustainability management and resource efficiency.

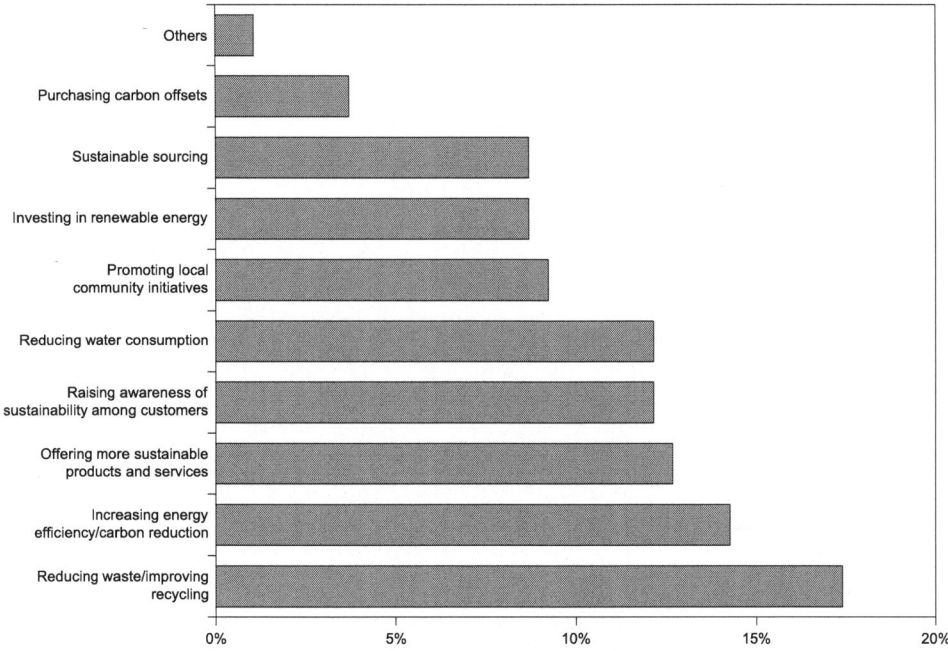

Figure 2.3: Resource efficiency

(Source: Data adapted from Murray, S [Gardner, B and Tabary, Z (eds)] (2013) *Sustainability Insights: Learning from Business Leaders*, A Coca-Cola Enterprises report, written by the Economist Intelligence Unit, 1 October 2013)

Coca-Cola Hellenic Bottling Company (Coca-Cola HBC) emits nearly 900 thousand tons of CO_2 per year as a result of consuming energy during manufacturing and running its transport fleet. The bottling of Coca-Cola™ also consumes food grade CO_2 purchased from external suppliers. Apart from the various initiatives to improve plant energy efficiency by 23 per cent, one of the key initiatives is for Coca-Cola HBC to reduce its emissions by generating energy efficiently using a tri-generation (heating, cooling and power) plant, and to recover CO_2 generated from the tri-generation plant for reuse in its bottling plant.

A tri-generation plant with CO_2 recovery is known in the industry as 'quad-generation'. With an investment of nearly €200 million, quad-generation plants were installed in Hungary, Romania, Northern Ireland, Italy, Nigeria, Poland and the Ukraine. Coca-Cola HBC estimates that this initiative will reduce 40 per cent of its direct emissions from energy generation and recover approximately 40 per cent of food grade CO_2 for use in its bottling plant.[54]

Supply chain efficiency

Apart from carrying out an energy audit within the boundaries of the organization, many organizations are auditing their supply chain as an opportunity to effect total energy reduction from their products and services. In an Economist Intelligence Unit study,[55] 61 per cent of managers acknowledge the significant contribution from supply chain collaboration and transparency. Furthermore, as shown in the various examples earlier, a significant amount of energy-related CO_2 emissions comes from within an organization's supply chain. In fact, research by the Cambridge Centre for Climate Change Mitigation Research (4CMR) suggests that up to 60 per cent of global energy-related CO_2 emissions appear in the supply chain of 500 corporations.[56]

As such, the same energy audit methodology or standard can be applied to an organization's supply chain. Table 2.2 shows how ISO 50002 can be used in resource efficiency and in supply chain efficiency initiatives.

As another example, GSK has a long-term vision for its own operations and for its supply chain to be carbon-neutral by 2050. An analysis of the CO_2 emissions of its products, from the raw materials to their final use by its customers, revealed that its own business operations and manufacturing plants only emitted 11 per cent of CO_2. Thirty-eight per cent came from CO_2 emitted from manufacturing its raw materials, 35 per cent of CO_2 was emitted from the use of medicinal inhalers, where CO_2 is the propellant, by its customers, 7 per cent from other end-customer use (e.g. boiling water to make Horlicks™), 3 per cent from logistics and business travel, and less than 1 per cent from disposal.[57]

Detailed knowledge of GSK's supply chain showed GSK that 89 per cent of the CO_2 arose not from its manufacturing processes but from elsewhere. If GSK was to focus purely on its own operations and become carbon-neutral, its products would still emit 13.4 million tons of CO_2 via its entire value chain. It also showed GSK that a significant quantity of CO_2 emissions came from the supply chain of two products: inhalers and Horlicks™. This allowed GSK to formulate strategies and prioritize R&D activities to maximize CO_2 reduction and minimize the impact of its

Chapter 2 Energy audit standards

products on the environment, e.g. low-carbon inhalers, recycling of its inhalers and working with its supply chain to minimize energy consumption, while maximizing its own operational efficiencies and energy reduction.

On a slightly different approach, Tesco has aspirations to reduce the carbon footprint of the products it sells by 30 per cent. It leveraged the collective know-how of its 700-plus suppliers and its own purchasing powers to help its supply chain reduce energy consumption.

Tesco first tested the initiative between February 2013 and September 2013, with four suppliers with a total floor space of 750,000 sq. ft. The trial involved using Tesco's buying power to install LED lighting for the four suppliers. The suppliers were able to reduce their lighting costs by 80 per cent (1 million kWh per year) and it cost them 25 per cent less than if they had invested on their own.[58] Following on from the successful trial, the initiative was extended to other members of Tesco's supply chain.

Different uses of energy audit standards

Table 2.2: Applying ISO 50002 to resource efficiency and supply chain efficiency initiatives

	Type 1	Type 2	Type 3
Resource efficiency	Auditing of a few key components as part of a larger sustainability/resource audit: energy, water, raw materials, improving yield, waste, packaging, transport, etc. High-level snapshot for medium- and large-sized organizations to identify and prioritize efforts. Savings are based on component improvement only. Carried out by equipment supplier or by specialist as a precursor to type 2 work. Utilizes 'rules of thumb' and 'benchmarks' in estimating savings and capital costs. Energy savings and capital costs are typically to an accuracy of ±50 per cent. Low-cost, low-risk opportunities are implemented at the end of this stage. Typically a half to one day for every component of audit.	Auditing of all resources on a manufacturing facility, site and/or building. Assesses the interaction and synergy of resources on the site or building. Audits carried out by competent people with knowledge and experience of various components, or by several teams, covering all the components of the audit. Savings are based on optimizing all resources in the whole system. Energy savings and capital costs are typically to an accuracy of ±25 per cent. Some opportunities are implemented at this stage. Others are progressed to type 3. Typically 5 to 10 days' site work or 2 to 5 days, using multiple teams.	Audit based on a whole building and/or site, taking into consideration the interaction and synergy of all resources. Typically used for a detailed audit of specific opportunities involving high business or capital risk. Audits are carried out by specialists with in-depth knowledge and experience, or a team of specialists, covering all the components of the audit. Savings are based on measurements and calculations to optimize all resources in the whole system. The calculation may be aided by computer simulation. Energy savings and capital costs are typically to an accuracy of ±10 per cent. Depending on the size, can range from one month to one year.

Chapter 2 Energy audit standards

	Type 1	Type 2	Type 3
Supply chain efficiency	High-level audit of every entity in the supply chain to collect data to access and quantify supply chain energy consumption or carbon footprint per unit of product and/or service. Heavily focused on desktop data analysis. Mostly led by one consultancy firm with technical understanding of the 'sponsor' organization, and supported by others with specific competencies and skills for the entity's industry sector or cluster. Low-cost, low-risk opportunities are implemented at the end of this stage. Typically a half to one day per entity in the supply chain.	Detailed audit of every entity in the supply chain with the aim of quantifying energy flows and identifying opportunities for improvement for each individual entity, with a view to minimizing the energy consumption of the whole supply chain. Detailed operating knowledge of the 'sponsor' organization is necessary to assess and optimize energy consumption in the supply chain. Type 3 audit is not normally carried out for a supply chain. Implementation of opportunities identified in the previous level(s) is either by the specific entity or jointly between the 'sponsor' organization and the specific entity. Mostly led by one consultancy firm with technical understanding of the 'sponsor' organization, and supported by others with specific competencies and skills for the entity's industry sector or cluster. Savings are based on measurements and calculations to optimize energy consumption in the overall supply chain. The calculation may be aided by computer simulation. Typically, 5 to 10 days of site work for each entity, with detailed supply chain analysis once all audits are complete.	

What does an energy audit not solve?

Figure 2.4 shows the results of the top 10 barriers business leaders and business managers said are stopping them from investing in energy efficiency.[59] Of these showstoppers, 70 per cent of them (failure to assess the side effects or consequences (of inaction), leadership attitude towards avoiding new costs, insufficient collaboration among stakeholders, corporate culture resistance to new ideas, and financial constraints, etc.) are within the grasp and direct influence of the company itself.

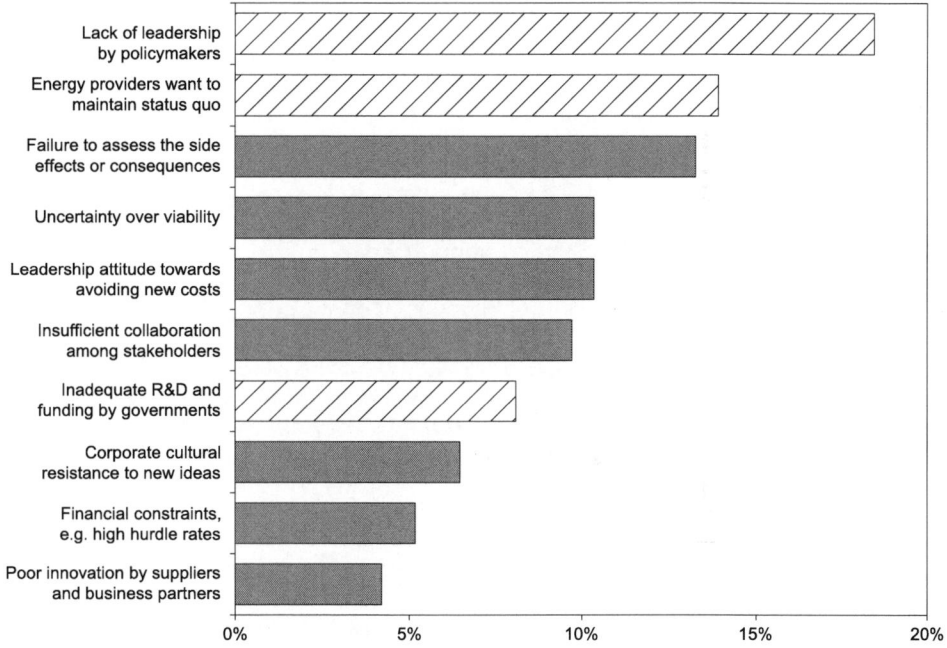

Figure 2.4: Top 10 barriers towards energy efficiency

(Source: Data adapted from Harvard Business Review Analytic Services (2013) *The Future of Energy*, A Harvard Business Review Analytic Services Report)

Issues like 'failure to assess the side effects or consequences' and 'uncertainty over viability' relate to how employees within an organization relate to and communicate with one another on the subject of energy reduction and energy efficiency. Energy reduction is normally expressed as a cost reduction initiative. At other times, especially when there are other benefits associated with the project, they too need to be taken into account and communicated. Some of these benefits might be

Chapter 2 Energy audit standards

improved product quality, reduction of rejects, improved work environment, reduced maintenance and reduced asset degradation. External energy auditors can help to identify these opportunities. However, they lack the insider knowledge, and hence need the organization to put a value to these additional benefits.

Organizations that require a high hurdle rate or short paybacks to investments are usually:

1) technically unsure of the economic viability of the project;
2) unaware of how the organization structures its costs and how debts impact on the cost of capital; or
3) unaware of alternative financing mechanisms.

Frequently, these short-termism views also rely on debt financing the business, and businesses find themselves cycling between periods of good income and struggling to stay afloat. Couple this with poor economic periods, and many lay off their employees and increasingly focus on quarterly profitability. To these companies, energy reduction and energy auditing is another cost and another project to take on.

These attitudes and aspects to business are primarily a European and American way of thinking and way of doing business. Many studies, including the latest published by Bain & Company,[60] show that organizations outside Europe and America, especially those in Brazil, Russia, India, China and South Africa (BRICS) focus on long-term results. Business models, innovation and cost structures are geared towards a steady, continual and sustainable growth. They also have a larger proportion of newly built building stock, plants and machinery that are inherently more energy efficient.

A good-quality, transparent and comprehensive energy audit can show the managers how they can reduce energy consumption and save energy costs. However, the report will not be able to influence the organization's people dynamics, finance structure and business strategy. The managers in these organizations need to learn the macroeconomics of their business and begin to look at longer-term business results and profitability.

Chapter 3 Understanding energy use, energy consumption and energy efficiency

As described in Chapter 1, the very first task of managing energy and finding opportunities to save energy is to understand where the organization uses energy. For many managers, this will sound like a daunting task, especially for non-science and engineering people; this chapter could get boring and this should be the remit of scientists and engineers. This is primarily because technical terms come into play. Sometimes, these terms sound similar or have confusing meanings.

Getting bored with the mechanics of energy consumption is the key barrier to generating significant value from saving energy. By increasing an understanding of how the organization uses energy, preconceived ideas and notions can melt and morph into creative, and sometimes innovative, ways to save energy. It starts by having the ability to ask better questions and gain insights from the existing equipment and systems.

Standards for energy management and energy audits define three specific terms relating to energy. They are 'energy use', 'energy consumption' and 'energy efficiency'.

'Energy use' refers to the application of energy by a user of energy. A simplistic way to identify energy use is to ask, 'Is this equipment connected to electricity, steam, hot water, compressed air, vacuum, chilled water, cooling water, glycol lines, natural gas and/or another fuel source?' Everything in a manufacturing plant, in offices, in commercial areas and in transport that needs energy to function is an energy user. Some examples are process equipment (reactors, distillation equipment, dryers, furnaces, etc.), utilities equipment (boilers, chillers, air compressors, cooling towers, etc.), heat exchangers (reboilers, condensers, evaporators, etc.), ventilation systems, lighting, motors, compressors, pumps, fans, IT equipment and transport (forklifts, cars, buses, etc.).

The second term is 'energy consumption' – the quantity of energy 'consumed' by the energy user for it to function. The measurement of energy is kilowatt-hour (kWh). Natural gas, LPG and fuel oils are normally measured in cubic metres (m^3) or litres (l). Solid fuels such as coal and solid waste are normally measured in kilograms (kg) or tons (t).

Chapter 3 Understanding energy use

The utility company converts these into energy units such as kWh, therms or British thermal units (Btu) for invoicing. For large organizations, and commercial and industrial companies, a breakdown of energy supplied to the organization can be obtained from the utility company.

Scientists and engineers would quote Sir Isaac Newton's law of thermodynamics where: energy cannot be created nor destroyed; it is transformed from one form to another. Therefore, the word 'consumption' is a wrong choice of word. Based on the laws of physics, the scientist and the engineer are right. Colloquially, and as defined by the various energy standards, energy consumption is used to mean: 'To use the portion of energy that is beneficial to that specific energy user.' The portion that is not used or cannot be used is deemed as a waste or a loss.

Of boilers, cooling towers, chillers, air compressors and ventilation systems

Organizations purchase energy in the form of natural gas, LPG, fuel oil and electricity. Some companies use this energy directly in their manufacturing processes. Most use energy indirectly by transforming the energy purchased into a different – and usable – form of energy: steam, hot water, compressed air, cooling water, chilled water, electricity (at a different voltage), etc. The 'transformed' energy could also be used directly in a process or by a piece of equipment, or could be further transformed, e.g. via a heat exchanger.

In industry and commerce, this 'transformed' energy is commonly known as a 'utility'. In some organizations, it may also be known as a 'facility'. In the building sector, it is commonly referred to as 'building services'. To add another layer of complexity to the term 'energy', fossil fuel, such as natural gas, LPG, fuel oil and coal, is called 'primary energy'. Organizations purchase primary energy to generate (or transform it into) secondary energy. A power plant consumes primary energy to generate electricity (a secondary energy).

The energy supply for some companies may come from a CHP plant, a central utilities plant servicing several companies, and those adjacent to another organization with excess power or heat may have a commercial agreement in place to purchase their excess energy or energy capacity.

Normally, organizations operate their utilities plant to match the highest-quality demand among all of the energy use in the organization. For example, steam could be generated at 10 barg pressure because the highest steam pressure user requires it at 9.5 barg. As another example, chilled water could be generated at 6 °C because the lowest temperature requirement in the organization is at 7 °C.

Steam boiler

A steam boiler works in tandem with its water treatment system to generate boiler quality make-up water. This make-up water, together with any condensate returned from the site, is collected in a hotwell tank (if the hotwell tank is pressurized, it is called a deaerator). From here, water is pumped by a boiler feed pump into the boiler to generate steam. Some boiler plants have an economizer to recover some heat from the hot boiler exhaust before going to the chimney. Some steam is used to maintain the water temperature in the hotwell tank and the remaining is distributed to the end users. Figure 3.1 shows the interrelationship between the steam boiler and the ancillary plants.

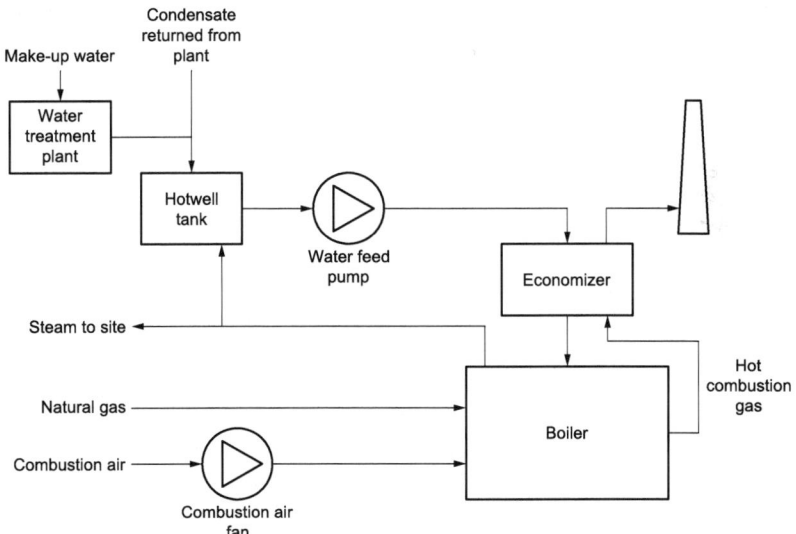

Figure 3.1: Steam boiler and ancillary plants

Chapter 3 Understanding energy use

Hot water, cooling water, chilled water and chilled glycol

A hot water boiler can have several configurations depending on the type of boiler, control philosophy and material of construction. Some boilers have an integral water storage facility and therefore do not need a hot water buffer tank. This type of boiler tends to be fully modulating, i.e. the pump can vary according to demand. Just like steam boilers, some hot water boilers have an economizer to recover heat from the hot exhaust to preheat the water before the main heat transfer takes place.

The three most common types of configuration are shown in Figure 3.2. Depending on the design, a hot water distribution system could have one or two buffer tanks (one for supply and one for return), or one or two separate main pipe headers for distribution.

All hot water systems will have pumps for distributing the hot water. The pump supplying the boiler is called the primary pump. The pump supplying the distribution is called the secondary pump. The hot water may be used directly by the end user or indirectly via a heat exchanger.

In large industrial, commercial or building blocks, it is also common to find secondary pumps supplying tertiary loops acting as local distribution nodes, and so on.

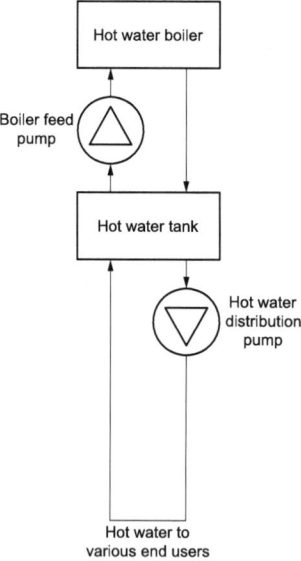

Figure 3.2a

Of boilers, cooling towers, chillers, air compressors and ventilation systems

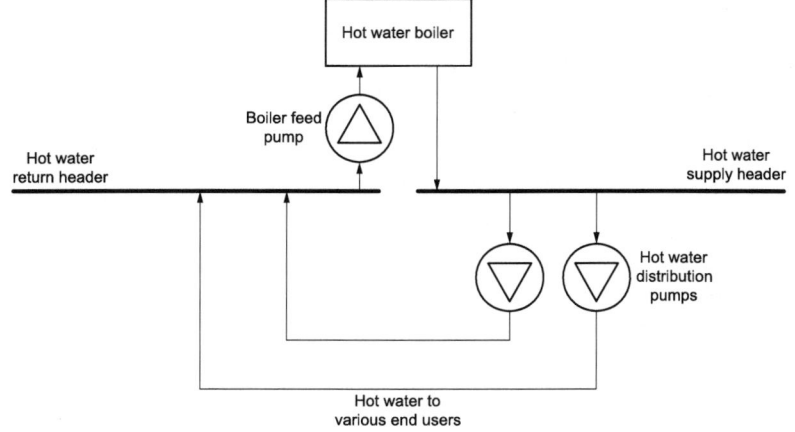

Figure 3.2b

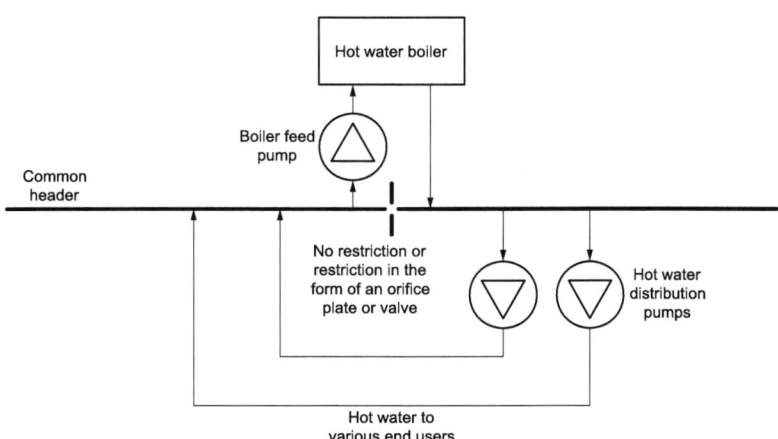

Figure 3.2c

Figure 3.2: Common configurations of hot water boilers and ancillary plants: (a) simple system (b) two separate headers configuration and (c) attemperation configuration

Cooling water, chilled water and chilled glycol also have similar distribution systems to that of a hot water system. Instead of a boiler adding heat into the water, a cooling tower or chiller is used to extract heat from the loop. The heat extraction system could be an air-cooled or a water-cooled tower.

Air compressor

Figure 3.3 shows the interrelationship of an air compressor with its ancillary plant. Ambient air contains moisture know as humidity. Air from the atmosphere is compressed in an air compressor. After compression, the air is hot. This is cooled down to an ambient temperature in an after cooler. The moisture in the compressed air and some entrained lubricant oil is then separated and drained in a water separator. Then, the compressed air is filtered to remove particulates. Depending on the air quality required by the end user, the compressed air is treated until it is in a usable form, before distribution. This treatment could take the form of using additional or varying grades of air filters, or moisture removal, to achieve additional levels of dryness. Some organizations may utilize air receivers as a buffer between the compressor and the air dryer. These are called 'wet receivers'. Other organizations may utilize air receivers between the air dryer and the end user (called 'dry receivers').

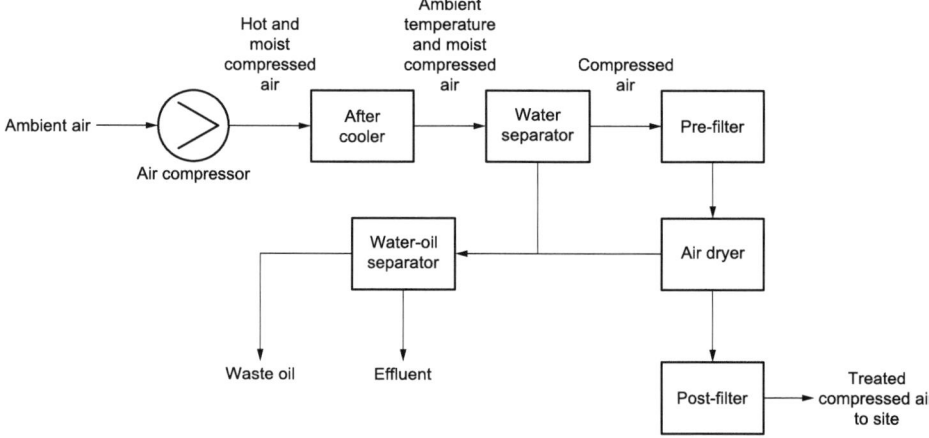

Figure 3.3: Air compressor and ancillary plant

Ventilation system

Another common energy use is that of the ventilation system. Within a building, ventilation can account for up to 60 per cent of energy consumption. Ventilation systems also have a wide variation in terms of size, types of installation, types of application, and combination of heating, cooling and air conditioning options. At its most basic form, a ventilation system consists of:

Of boilers, cooling towers, chillers, air compressors and ventilation systems

1) a preheating coil to preheat air if the ambient temperature is below freezing temperature;
2) a series of pre-filters to remove any particulates to meet user requirements;
3) a fan that provides the air movement;
4) additional heating, cooling, humidification or dehumidification of the air; and
5) another series of filters to suit user requirements.

Depending on the type of installation, if all ventilation is supplied by 100 per cent outside air, this is known as a 'full fresh' system. The supplied air leaks out through the doors, windows and any other opening. In other systems, a fan is used to extract 'stale' air from the end use. Where an extract fan is utilized, it may contain a series of filters to filter out any contamination from the end use. These may be in the form of particles or gases.

Some ventilation systems recycle the air back into the suction side of the ventilation system by way of an additional fan and damper arrangement. These ventilation systems are known as 'recirculation' systems. Some other ventilation systems extract heat from the hotter extract air and uses this to preheat the cooler incoming fresh air. Figure 3.4 shows an example of a ventilation system.

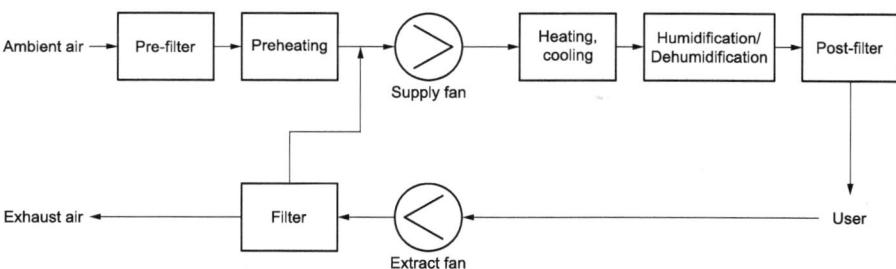

Figure 3.4: Example of a typical ventilation system

The heating, cooling, humidification and/or dehumidification functions may be supplied by an external source of energy such as a boiler, chiller, humidifier or dehumidifier. These functions may also be self-contained within the ventilation system.

In a large building, as described above, the ventilation could be a centrally located unit supplying air to a 'service shaft' and each floor consumes air from this 'service shaft'. The ventilation system could be mounted on the ceiling or above a window. Very often, for ceiling or window mounted ventilation systems, the system is 'split' into two parts:

Chapter 3 Understanding energy use

the ventilation fan inside the room and the rest located outside. The building may also have a hybrid ventilation system where a central ventilation system services several smaller, ceiling-mounted ventilation units.

A machine can be energy efficient and still use more energy

It is very common for organizations to assume that they consume 100 per cent of the energy they buy. In practice, every time energy is 'transformed' or 'converted from one form of energy to another', a fraction of its energy is 'lost'.

When natural gas is burned in a boiler to generate steam or hot water, some energy loss occurs. When electricity is used in an air compressor to generate compressed air, some energy loss occurs and leaves the compressor as heat. This gives rise to the third definition used in energy standards: 'energy efficiency'. This is sometimes called the 'fifth fuel' or the 'invisible fuel'.[61]

Scientifically, energy efficiency is:

$$\text{Energy efficiency} = \frac{\text{Total energy input - unused energy}}{\text{Total energy input}} \times 100$$

Taking away from the 'total energy input' any energy that was 'unused' or 'lost' gives the quantity of energy that was consumed. A simplification of the equation above becomes:

$$\text{Energy efficiency} = \frac{\text{Useful energy output}}{\text{Total energy input}} \times 100$$

It follows that the bigger the energy losses, the more energy the organization needs to buy to meet its demand. If a steam boiler is 80 per cent efficient, the company needs to buy natural gas equivalent to 100 kWh to generate 80 kWh of steam. Managers tend to use the energy efficiency figure for budgeting purposes. For example, if an organization is manufacturing 1,000 units of product and each unit consumes 10 kWh of steam, the organization needs to purchase:

$$\frac{1{,}000 \text{ unit of product} \times 10 \text{ kWh of steam}}{80\ \%\ \text{boiler efficiency}} = 12{,}500 \text{ kWh natural gas}$$

A machine can be energy efficient and still use more energy

Although the use of boiler efficiency for budgeting is correct, it hides energy waste or energy loss from the conscious mind: for example, is 80 per cent the best at which the said boiler can operate? What is the 'best in class' boiler efficiency at the specified load conditions? What opportunities are available to improve it beyond 80 per cent?

Over time, if the '80 per cent' figure is not challenged, people in the organization begin to engrain it in their minds, use it as the de facto benchmark and forget that there may be opportunities to improve it. Defining energy efficiency as intended – scientifically – reminds people that energy could be wasted and/or lost, and reinforces the need to use it effectively. In the case of the 80 per cent-efficient boiler, there are possibilities to increase the boiler efficiency beyond 80 per cent.

An organization, using the concept of energy efficiency, can assess the efficiency and effectiveness of its steam, hot water, cooling water, chilled water, chilled glycol and compressed air generation systems. Other areas that are also commonly audited are electricity distribution, power quality correction, voltage reduction, lighting improvements, controls in air handling units, and the application of VSDs for motors, pumps and fans.

One of the drawbacks of a single equipment energy audit (or installing the most energy-efficient machine) is that the energy retrofit can be the most energy efficient, but the efficiency of the system does not improve. The following are some examples where energy-efficient retrofits do not produce energy savings or energy savings are not maximized:

- the most commonly found example is installing VSDs for fixed-speed applications and where bypass valves or throttling valves have not been removed. The purpose of a VSD is to vary the speed of the motor to match the demands of the pump or fan. As there is no signal to tell the VSD to slow down or speed up, installing a VSD may not maximize the available energy reduction;
- many organizations install high-efficiency condensing boilers but continue to operate the boilers above 60 °C. When the supply temperature is well above 60 °C, the condensing boilers are not condensing, and the energy reduction benefit from the higher efficiency is not realized. The benefits of a condensing boiler are maximized by operating the hot water system at lower temperatures, e.g. for underfloor heating;
- voltage reduction works on lighting with non-electronic ballast and small power systems. For other plant machinery and rotating machinery, such as pumps, fans and motors, a reduction in voltage will result in an increase in electricity consumption;
- some air compressor manufacturers market their products as 'carbon-neutral'. Organizations may purchase and install them thinking that they will be the most efficient. Careful inspection of the claims shows that the air compressor is carbon-neutral under specified conditions, e.g. when operated at a location equivalent to

Energy Audits

Chapter 3 Understanding energy use

sea level, taking in ambient air at 40 °C, compressing the air and finally cooling the air to 20 °C by generating hot water at 70 °C to 90 °C. This means that if any of the conditions are not meeting the specified conditions, the compressor will not be carbon-neutral but will consume energy;
- another commonly found example is installing high-efficiency (also known as coefficient of performance (COP) or Energy Efficiency Ratio (EER)) chillers with poorly designed distribution systems, i.e. a large bypass, attemperation loops, three-way bypass valves or throttling valves. A large proportion of the chilled water does not go to and/or through the end user. The cumulative effect of the large bypass is that the chiller is operated at low loads and becomes energy inefficient.

Identifying energy use and variables that cause consumption to vary

Following on from the previous section, it is important to understand what can cause energy consumption and energy efficiency to vary. A good tool to identify this valuable energy information is the use of success maps. This technique, originally developed by Robert Kaplan and David Norton as 'strategy maps', asks the question: 'What can cause the energy consumption of this equipment to vary?'

Success maps – tool to identify variables

Let us use a building's heating system as an example. To develop a success map for the heating's energy use, i.e. of natural gas, ask and answer the following:

Question: What can cause the natural gas consumption of the boiler to change?

Answer:

1) the room temperature is too high;
2) the heating system has become inefficient;
3) the cooling water valve is not fully closing; or
4) there is a cooler ambient air temperature.

A similar question-and-answer technique can be used to drill down the causes and reach the human interface level. Using the same example of a building's heating system:

Question: What can cause the heating system to become inefficient?

Identifying energy use and variables that cause consumption to vary

Answer:

1) the temperature probe is faulty;
2) there is missing or faulty insulation; or
3) there is scaling on hot surfaces.

An example of a completed success map for the above example is in Figure 3.5. This success mapping technique is equally applicable in the manufacturing environment. Figure 3.6 shows an example of an air dryer. A series of these maps can be developed for the whole organization to give an overall picture of energy users.

Success maps create visibility of energy use and energy consumption within the organization. For many, it is the first step towards making energy consumption a variable cost. Creating and using success maps can give the following information:

1) *there are many sources of energy use and energy consumption information in the organization.* A success map identifies all of the energy and energy-related information that is already available in the organization. Many measurements are already available. Some examples are flow, weight, volume, temperature, pressure, time 'on' or time 'off' status, number of push buttons, composition, pH, conductivity and power factor – information collated from local gauges, lab testing, supervisory control and data acquisition (SCADA) systems, human–machine interfaces (HMI), distributed control systems (DCSs), building management systems (BMSs) and/or energy information management systems (EIMSs)[5]. This is the information the organization uses for day-to-day operations and it can be converted into energy information. This valuable operational information can be used to calculate energy consumption and energy efficiency information;
2) *identify which parameters are important to energy consumption.* The success map highlights all the parameters that change the quantum of energy consumption for specific equipment and the system. They need to be measured and controlled to manage energy efficiently. If there are parameters that are not measured by existing meters, new meters should be installed to measure them;
3) *different groups of people relate to, and use, different energy-related information.* It shows that different people within the organization use different information for their day-to-day work. In an organization, operators and technicians will readily relate to information such as flows, temperatures and pressures. However, a manager in the same organization will relate to other parameters, depending on their function within the business. Energy managers

---

[5] An energy information management system (EIMS) is similar to a BMS. Some manufacturers call their EIMS simply an 'energy management system'. These are not to be confused with ISO 50001-based energy management systems.

will relate to energy parameters, production managers will relate to production throughput, etc. To communicate effectively with different groups of people, it is important to tailor different sets of information to the needs of the people;

4) *different groups of people need different sets of energy performance indicators.* As different groups of people within the organization use different sets of parameters, they also need different sets of energy performance indicators to one another. Operators, maintenance technicians and front-end office staff need indicators that are local to their work environment. Examples are temperature and flow. A functional manager needs an indicator that is representative of their function. Examples are boiler efficiency for the utilities manager and energy consumption during idle time for the manufacturing manager. For the senior managers, an overall company indicator is more suitable. Using one energy performance indicator for everyone in the company means that many employees cannot relate to the information and are less able to participate in the company's efforts to reduce energy consumption.

Armed with the knowledge of which equipment or machine uses how much energy and an understanding of what causes it to vary, the organization can begin to view energy as a variable cost (as opposed to a fixed cost), and bring visibility to how energy is used, and how it could be effectively managed and controlled.

When organizations realize they have more 'energy users' than originally understood, there is a tendency to rush into the installation of additional energy meters. As the saying goes, 'you can't control what you don't measure'.

Meters, meters everywhere

The need for energy information does not necessarily mean that organizations should buy and install energy meters and automated data logging systems in every location. Doing so can be costly and swamp the organization with information that may be confusing or have little use apart from monthly energy reporting and budgeting. In the end, these meters may become a white elephant.

A good place to start is to overlay sources of energy information onto the success map. Some of the information may be spot measurements; other measurements could be calculated based on known information. Once this process is complete, the success map can also function as a 'metering plan', with which the organization can decide what variables are not to be measured and where additional metering would be beneficial.

Identifying energy use and variables that cause consumption to vary

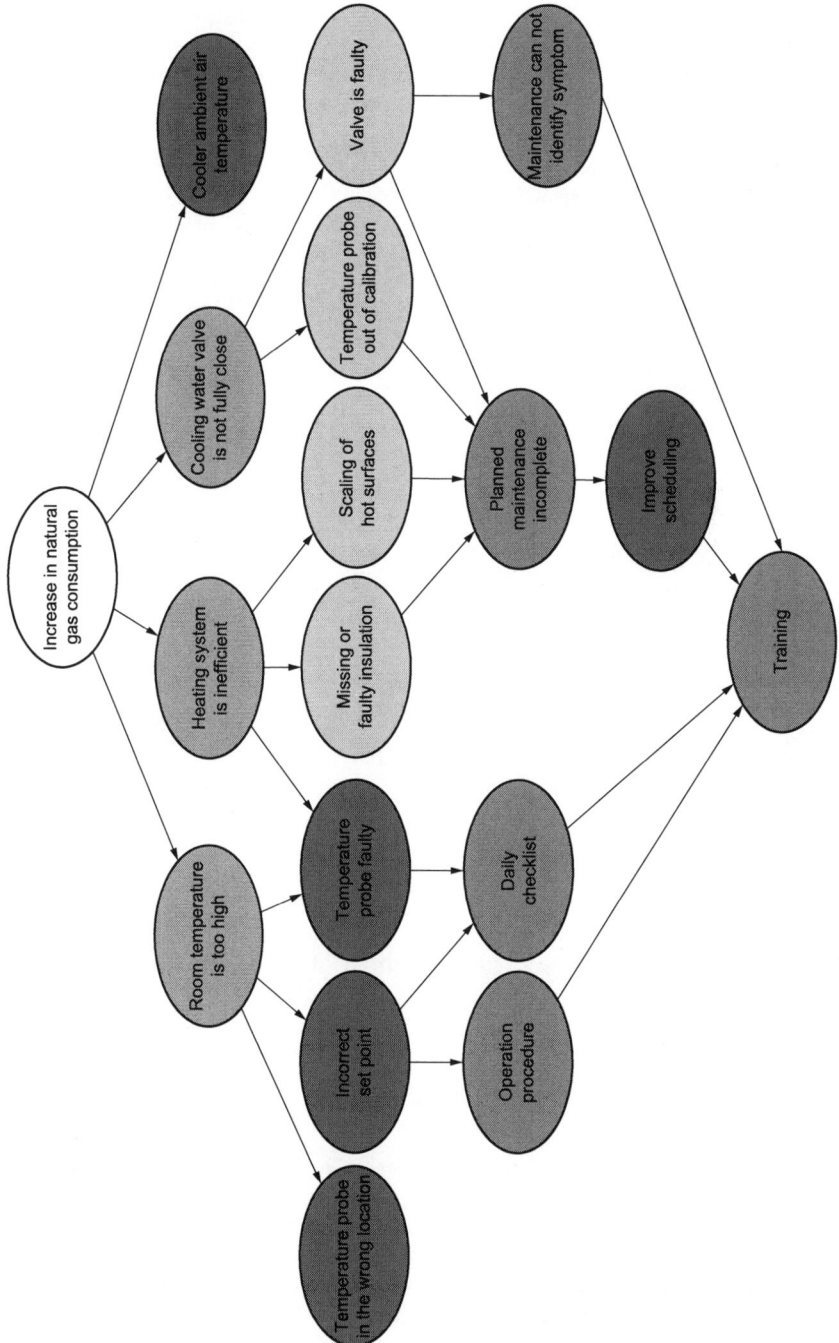

Figure 3.5: Success map for an office heating system

Chapter 3 Understanding energy use

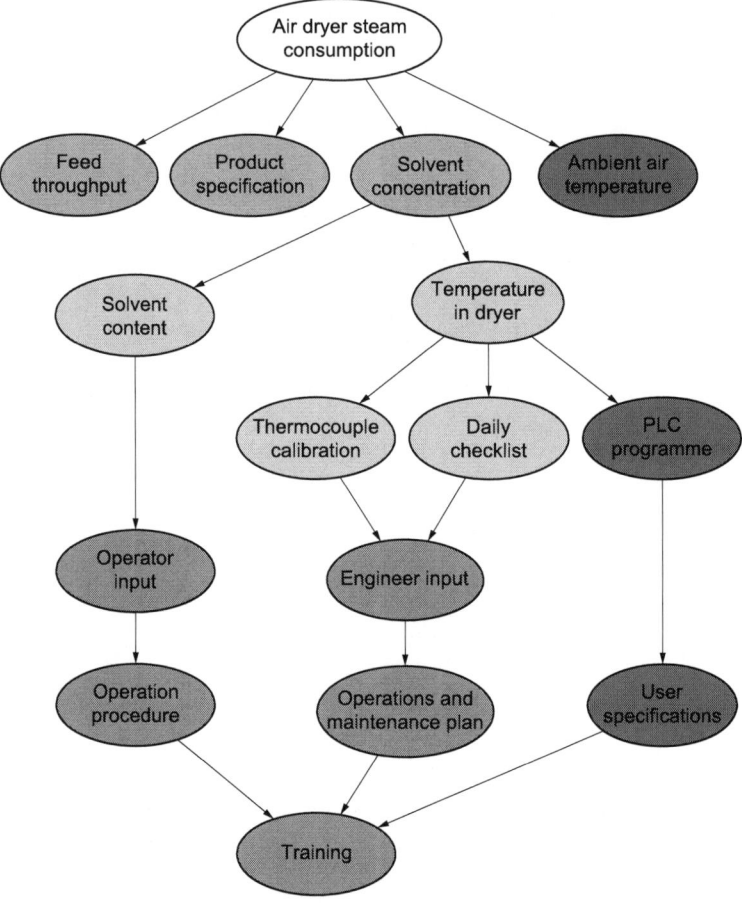

Figure 3.6: Success map for an air dryer in a manufacturing plant

(Source: Oung, K (2013) *Energy Management in Business*. Farnham: Gower)

There are two major and common variables that affect energy consumption. These are:

- production;
- weather.

A good understanding of these variables is necessary to build insight into how the organization consumes energy, and how it means to control and manage this resource.

Production as a variable

Figure 3.7 shows a plot of energy consumption versus production. A best-fit line can be drawn on the data with a straight-line relationship:

$Y = mX + c$

Energy consumption = Slope × Production + Constant

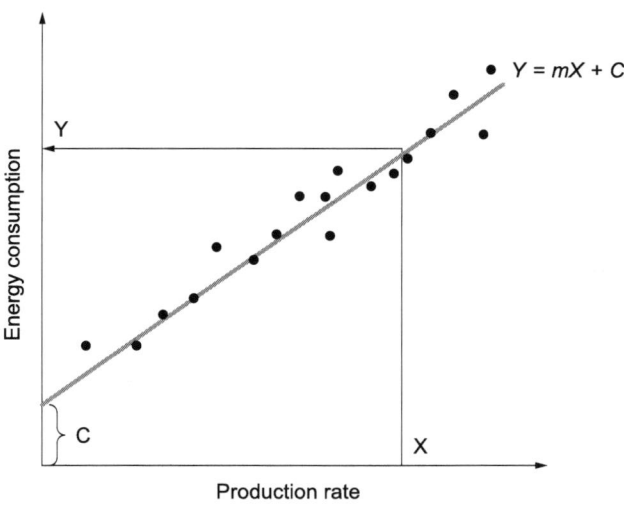

Figure 3.7: Production as an energy variable

In general, there are four types of information that can be extracted from this energy consumption–production relationship. This book explores three of the four types of information. For a good description of the various types of information that can be extracted and the various forms of energy consumption–production relationships, please see *Energy Management in Business: The Manager's Guide to Maximising and Sustaining Energy Reduction* (Oung, 2013) or any other good energy management books.

First, this relationship can be used to predict and budget for energy consumption. With a known production (X), the equation will estimate the expected energy consumption (Y).

Secondly, the quantity C denotes the energy consumption unrelated to production, i.e. the fixed load. A large C indicates that a large quantity of energy is not used to manufacture products and that the marginal energy consumption is small. A large C also gives the organization an opportunity to look for opportunities to reduce energy consumption

Chapter 3 Understanding energy use

without affecting production, i.e. machines idle for significant periods of time, energy being wasted, ventilation systems, laboratories, lighting, staff facilities, kitchens and offices, etc. Other fixed loads may be business critical and/or cannot be turned off or turned down. Some examples are data centres, server rooms and uninterruptible power supply rooms. A small C indicates the reverse. It is very rare for a manufacturing organization to have no C value in its plot of energy consumption versus production.

The third type of information is the slope of this relationship (m), the variable energy consumption, also known as 'marginal energy consumption' or 'specific energy consumption' (SEC). SEC is frequently used in benchmarking exercises and is quoted by licensed processes and when reporting energy intensities. Care should be taken when using SEC as a simple ratio as it can lead to errors in calculation:

- *computing SEC as a ratio of total energy consumption and good production output* has led to a lot of misleading plant performance information. A better way to calculate SEC is as the ratio of total energy consumption and total production output. This is because, if the bad-quality product is not detected until the end of a production run, a similar quantity of energy will have been consumed to manufacture the bad-quality product;
- *forgetting that SEC is the marginal or variable energy consumption and that there is a fixed component of energy consumption.* As mentioned earlier, it is rare for a manufacturing plant or building not to have a baseload energy consumption. A result of using SEC would indicate that when there is no production or building activity, there is no energy consumption. Therefore, SECs are only valid when the production output, or its range, is also specified.

In manufacturing organizations, not many energy consumption–production relationships conform to a straight-line relationship. Energy consumption versus production relationships other than a straight-line one are possible. For example, an organization that has a large population of rotating machines, such as power plants, rotary kilns, motors, pumps, fans, chillers, drives and air compressors, may exhibit a cubic relationship.

Some manufacturing organizations use the same machines to manufacture different products. Others may have a combination of weather and production variables. These components should be established independent of each other and then brought together into a multivariable (sometimes known as multivariate) analysis.

Identifying energy use and variables that cause consumption to vary

Weather as a variable

Figure 3.8 compares the weather profile of Worthing, UK and Kuala Lumpur, Malaysia across one year. As can be seen, there are different weather patterns across the year and also seasonal, weekly and even daily variations in the same location. A building or process and, hence, its energy consumption, will also vary with the weather profiles and weather extremes.

Similar to the energy consumption versus production curve, a best-fit line can be drawn on the data with a straight-line relationship:

$Y = mX + c$

Energy consumption = Slope × Heating degree day + Constant

Chapter 3 Understanding energy use

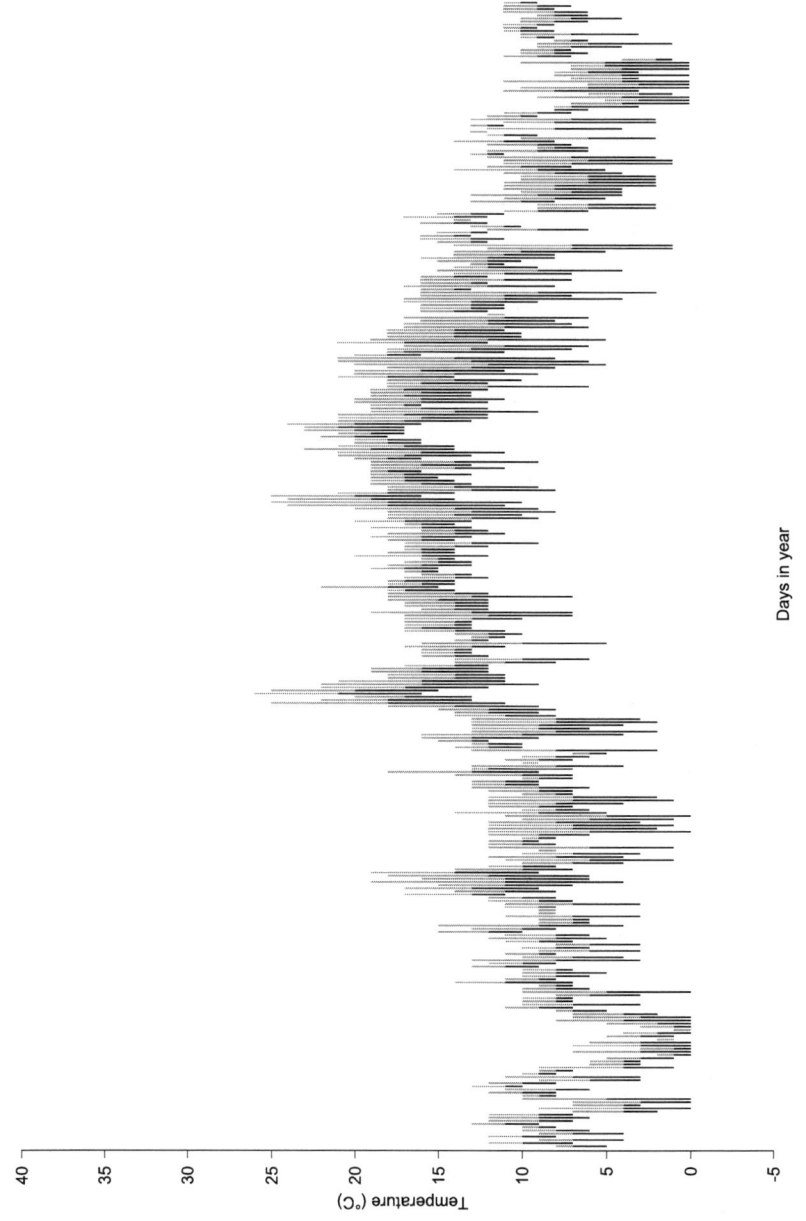

Figure 3.8a: Worthing

Identifying energy use and variables that cause consumption to vary

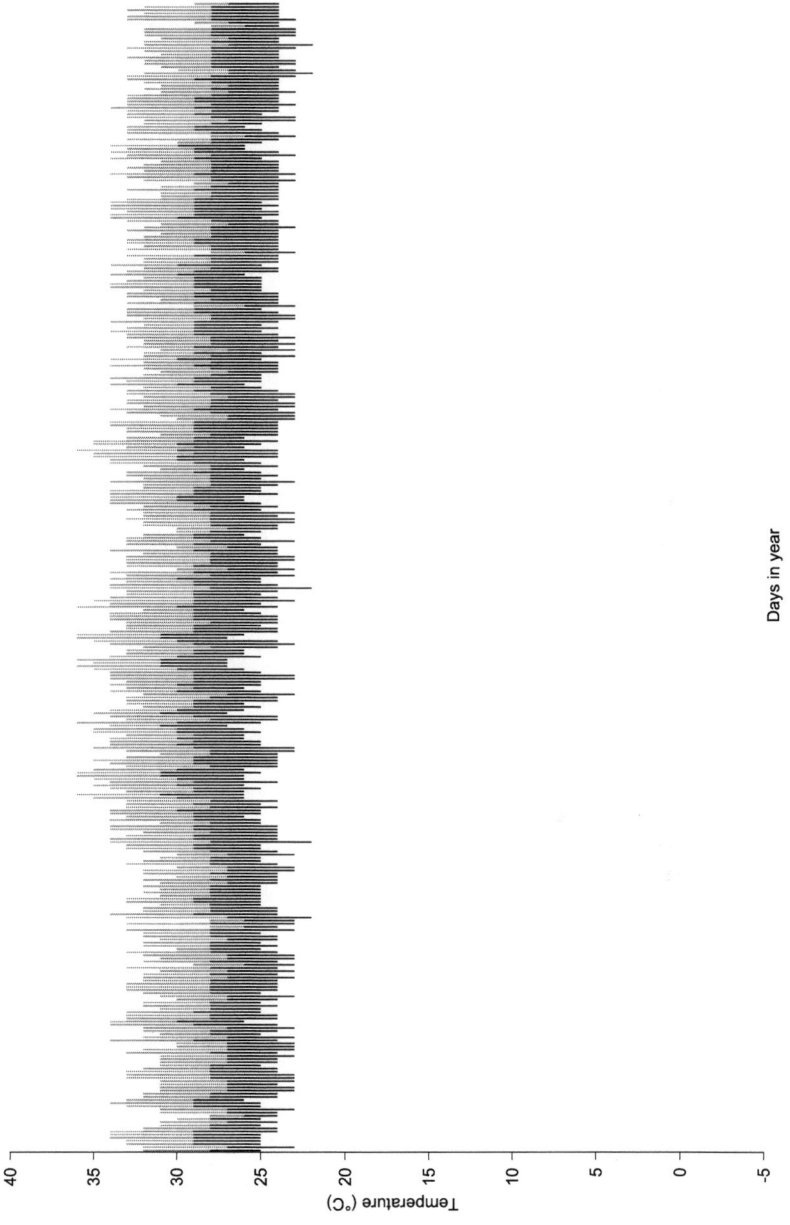

Figure 3.8b: Kuala Lumpur

Figure 3.8: Comparison of weather profile in 2012 of (a) Worthing, UK and (b) Kuala Lumpur, Malaysia

Chapter 3 Understanding energy use

Many heating degree-days' data with varying base temperatures are readily available from the internet and are used when heating is a variable for a building or a manufacturing process. The importance is to match the base temperature with the actual heating temperature set point for the building or process.

Again, similar to the energy consumption versus production relationship, the energy consumption versus weather relationship can be used to predict and budget for energy consumption. With a known heating degree-day (X), the equation can estimate the expected energy consumption (Y). Secondly, the quantity 'C' denotes the energy consumption unrelated to weather. A large C indicates that a large quantity of energy is not related to weather.

There are other weather data similar to heating degree-days, such as cooling degree-days and humidity-days. They are also available from the internet. With modern computers, it may be possible for an organization to compute the effects of weather on energy consumption using raw historical data, with more accuracy. This can be achieved using the building heat loss coefficient formula available on the internet.

To illustrate the importance of understanding the influence of weather on energy, let's have a look at the design of a data centre. Ventilation and air conditioning for data centres and computer rooms are normally installed using a close loop cooling system. This means that the hot extract air, typically at 24 °C, is extracted from the room, cooled to 18 °C using an air-cooled chiller, and returned into the room.

If the data centre is located in Kuala Lumpur, Malaysia, where the temperature is more than 18 °C for 100 per cent of the year, a close loop cooling system continually cools the air from 24 °C to 18 °C, or provides 6 °C of cooling. If the data centre uses fresh air cooling, it would be required to do more than 6 °C of cooling and therefore consume more energy.

In the case of Facebook's Prineville Data Center in Portland, Oregon, the temperature does not exceed 18 °C for 79 per cent of the year. This gives an opportunity for Facebook to utilize free air cooling, i.e. using fresh air to cool the data centre. For the remaining 21 per cent of the year, a chiller may be required. A lower-cost option, as used by Facebook, is adiabatic cooling, e.g. using water spray nozzles or demister pads. Obviously, when the ambient temperature is hotter than 24 °C, it makes sense to recycle the data centre air to minimize energy consumption. This method of cooling allows Facebook to reduce its energy consumption by 60 per cent.[62]

With Facebook's understanding of mechanical and electrical systems, coupled with its knowledge of computer and server technology, it was able to work with its computer, mechanical and electrical designers to minimize energy consumption and achieve a power usage effectiveness (PUE) of 1.08 versus an industry average of 1.8 and the 'best in class' of 1.2. PUE is a measure of how much energy is utilized in powering non-IT related energy use. A PUE of 1 means all energy purchased by the data centre is used to power the IT systems. For Facebook Prineville it means that 92 per cent of its electricity purchased is consumed to power the IT systems versus the average of 57 per cent. The data centre is, overall, 38 per cent more efficient and 24 per cent cheaper to build.[63]

Energy performance indicators and energy baselines

The next logical step, having gone through the process of identifying energy use, the quantity of energy consumed, and variables that can lead to a change in energy consumption, is to use this information to measure energy performance, use it as a means to control, manage and budget for energy consumption, and then to identify opportunities for improvement. Energy efficiency, as described earlier, is one such measure of energy performance.

Energy efficiency as a measure of energy performance

Also indicated in an earlier section, energy efficiency is a figure calculation based on the 'total energy input', 'unused energy' and energy consumed formula. For every energy user, the energy efficiency of said energy user can be calculated with any two known variables.

If a hot water boiler consumes 100 kW of natural gas and produces 75 kW equivalent of hot water, its efficiency is 75 per cent and 25 per cent of the energy is lost or wasted. If a ventilation fan consumes 10 kW of electricity and produces 4.5 kW equivalent of actual ventilation, its efficiency is 45 per cent and 55 per cent of the energy is lost.

Organizations can measure the relevant parameters and calculate energy efficiency. For example, by using a steam meter and a gas meter, the quantity of steam generated and natural gas used can be measured to calculate boiler efficiency. More often than not, the energy efficiency of equipment is back-calculated or inferred by a flue gas analyser. The analyser measures the oxygen, carbon monoxide and carbon dioxide content of the flue gas, and the flue gas temperature, to calculate boiler efficiency. Many other methods are also available.

Similarly, the efficiency of a chiller can be calculated by the chilling duty provided, i.e. by measuring the flow, supply and return temperatures into the chiller and the power consumed by the chiller. It can also be calculated by measuring the heat rejection from the chiller, i.e. by

Chapter 3 Understanding energy use

measuring the flow, supply and return temperatures from the chiller to the cooling tower and the power consumed by the chiller.

If an organization uses energy efficiency purely to compare the same energy use from one period to another period, and operates and maintains the same energy use, the use of energy efficiency as an indicator of energy performance will be adequate. However, if energy efficiency is to be used to compare energy performance with similar equipment located in a different location (building, site, region or country), or with 'best in class', several other parameters come into play. In general, the differences could come from six categories:

1) *gross and net efficiency.* The energy content contained in fossil fuels and other combustible materials is characterized by gross calorific values (higher heating values) and net calorific values (lower heating values). This energy is released when burned in an energy generation plant, i.e. boilers, furnaces and power plants. Depending on the fuel burned, the difference between energy efficiency calculated from gross calorific values and that calculated from net calorific values can be up to 16 per cent;

2) *primary versus delivered.* As described earlier in this chapter, when fossil fuel and other combustible material is burned, the energy released is said to be 'primary energy'. Delivered energy is the energy content as delivered to the end user. According to the laws of physics, every time a form of energy is converted or transformed, some energy content is lost. In some building or manufacturing plants, steam is further converted to hot water, and in that case energy conversion occurs four times:
   a) the boiler generates steam;
   b) the steam is delivered to the heat exchanger;
   c) the heat exchanger converts steam into hot water; and
   d) the hot water is delivered to the end user.
   While each conversion can be efficient, say 80 per cent boiler efficiency, 95 per cent steam distribution efficiency, 95 per cent heat exchanger efficiency and 95 per cent hot water distribution efficiency, the overall efficiency will be low, i.e. 80 per cent × 95 per cent × 95 per cent × 95 per cent = 68.6 per cent!

3) *different temperatures, pressures, humidity and sea level.* Many equipment suppliers define the energy efficiency of their equipment at a specific condition. Some examples are ambient temperature, ambient pressure, sea level and humidity. The difference in a steam boiler's efficiency calculated using a hot water temperature feed at 80 °C and at 150 °C could be up to 6 per cent;

4) *instantaneous versus average.* All equipment can operate within a range of demand. For example, a hot water boiler may operate from 25 per cent to 100 per cent of its capacity. However, the energy efficiency of said boiler will not be constant between 25 per cent and 100 per cent of its rated capacity. Using instantaneous or spot

measurements will result in an energy efficiency for that particular operating load. Measuring energy efficiency over a range of operating conditions and taking an average give a better representation of the machine when it operates;

5) *technical versus cost.* Some opportunities, such as a heat pump producing heat with a COP of three, can be highly energy efficient. However, if the electricity cost is four times more than natural gas, it would be cheaper to operate the heating using natural gas. As another example, an absorption chiller has a COP of 0.7. A mechanical chiller has a COP of 4. On paper, a mechanical chiller is more energy efficient. However, if there is a lot of low-grade heat available in the organization, e.g. by-products from a manufacturing process, recovering the waste heat to generate chilled water avoids the need to operate a mechanical chiller;

6) *composite ratings.* There is an increased use of 'seasonal efficiency' and 'integrated part load efficiency'. These are composite efficiencies using a reference location to calculate energy efficiency and take into consideration the regional weather profile and temperature range. Care should be taken when comparing composite efficiencies because different locations have different weather profiles. As shown earlier, Kuala Lumpur, Malaysia, has a significantly different weather profile to Worthing, UK. Using the energy performance data generated for a different location would lead to errors in calculation.

When assessing energy efficiency as applied to existing equipment and its potential for improvement, a consistent definition for energy efficiency is essential. As such, carrying out energy efficiency testing according to codes of practice and standards allows the efficiency of an energy user to be tested and compared on a like-for-like basis. A shortlist of common testing standards for energy efficiency is in Appendix B.

Energy performance indicators other than energy efficiency

As suggested in Chapters 1 and 2, although the concept of energy efficiency shows how efficiently a machine or a piece of equipment consumes energy, some energy use can be very efficient but does little or no useful work. For example, LED lighting is the most energy efficient lighting available. However, the LED lighting may be ON inside a building with good daylight where the lighting could be turned OFF without affecting the lighting levels. In such cases, the energy use is very energy efficient but is an unnecessary use.

As another example, a heating radiator may be the most energy efficient design. However, it is totally concealed in a wooden cupboard. As such, the radiator is efficient but ineffective in heating up the room.

In order to give energy information that is relevant to the various groups of people or departments, and which will solicit an action (or reaction) to

Chapter 3 Understanding energy use

reduce and minimize energy consumption, a set of information apart from energy efficiency is required – one that gives information relevant to these groups' or departments' job functions and areas of responsibility. As shown previously, a success map shows how different groups or departments within an organization relate to energy use, energy consumption and energy efficiency. Some examples of energy performance indicators other than energy efficiency are listed in Table 3.1.

Table 3.1: Examples of energy-related performance indicators for different job functions

Overall organization	Total energy consumption, energy consumption by type of energy, energy consumption by department, etc.
Manufacturing/services	Energy consumption per unit product, energy consumption per manufacturing line, energy consumption of manufacturing and non-value-added processes, energy consumption per unit of service, expected energy consumption, actual energy consumption versus budgeted energy consumption, cumulative sum (Cusum analysis, league tables, flow, temperature, pressure, level, defects, wastes, outstanding quality investigations, production scheduling and planning, etc.)
Maintenance	Availability, idle time, mean time between breakdowns, energy consumption during idle time, league table of same item maintenance, energy baseload when not in production or in service, number of outstanding maintenance tasks (e.g. number of steam traps in line for maintenance, or number of leaks), outstanding equipment failure investigations, etc.
Laboratory	Air change rate in laboratory, energy consumption per product or service rendered, fume cupboard sash height, equipment left 'on' when not in use, temperature, occupancy, etc.
Utilities	Generation efficiency, distribution efficiency, kW per unit utility, (e.g. 750 kW per ton of steam, or 0.1 kW/Nm^3 compressed air), COP, utility system efficiency, etc.
Purchasing	Purchasing according to energy-efficient specification, assessing energy performance as part of selection criteria, delivery of parts and equipment on time in full, etc.
Sales and marketing	Capacity utilization

Overall organization	Total energy consumption, energy consumption by type of energy, energy consumption by department, etc.
Human resources	Induction and refresher includes energy-efficient behaviour, employee appraisal includes items to encourage and motivate energy-efficient behaviour, ensure all personnel are competent to perform the required tasks, number of standard operating procedures under review, training records, number of energy suggestions, number of suggestions implemented, etc.
Office personnel	Air change rate, air flow rate, office temperature, office humidity, adherence to energy efficiency charter, reporting of maintenance issues, idle machines or equipment, occupancy rates, energy consumption per unit of floor space, etc.
Transport services	Litres of fuel per passenger miles, litres of fuel per kilogram of goods and miles travelled, etc.
Energy-related sustainability indicators	Water consumption, CO_2 emissions, supply chain miles, waste generation rates, waste reduction, recycle quantities, etc.

Modern day thinking about energy reduction

Traditionally, when organizations want to reduce energy consumption or improve their energy efficiency, activities are limited to identifying and implementing opportunities to improve the utility generation plants, electricity distribution, voltage reduction, lighting and air handling plants. A modern approach to energy efficiency assesses energy use, energy consumption and energy efficiency on a:

- system efficiency basis;
- resource efficiency basis;
- supply chain efficiency basis.

System efficiency

Picking up from the fact that (1) machines rarely operate independently from other machines and (2) a machine can be the most efficient, but the system as a whole can be inefficient, once the energy user and variable that causes it to vary has been identified, the organization can begin to assess the system's efficiency.

Chapter 3 Understanding energy use

Existing plant or building

A system is a set of machines or equipment that is interconnected via pipes or cables and is operated in a way that achieves its intended function. Analysing energy use as a system recognizes that there is a group of equipment or machines interacting with the other machines and also with humans. They do not exist in isolation but are interdependent and interlink via a series of pipes and electric cables. Calculating systems energy efficiency utilizes the same energy efficiency formula but takes into account auxiliary equipment, distribution losses and end use losses.

To take into account auxiliary equipment, it is simply referred to by adding the word 'house' in-between the main energy generation equipment and the word 'efficiency'. For example, 'boiler house efficiency' or 'compressor house efficiency'. The calculation is achieved by broadening the boundaries of the utility generation plant to include all of the auxiliary equipment and the interconnecting pipework, adding all of the energy inputs from the pumps, etc. and subtracting the energy content of the utilities consumed by the auxiliary equipment. This number will always be a smaller number than the energy efficiency for the utility generation plant alone.

Sometimes, the utility generation plant and its auxiliary equipment is very efficient but the losses occur elsewhere, i.e. in the distribution system or by the end user. There are four common reasons for losses in the distribution system:

1) leaks;
2) heat loss/heat gain;
3) the energy user;
4) system losses.

One of the reasons for energy loss in the distribution system is leaks. For example, boilers and compressors typically operate to a pressure set point located in the distribution system. When an energy user consumes steam or compressed air, the pressure in the distribution system drops, signalling the boiler or compressor to operate and make up the pressure inside the distribution system. A boiler, a compressor and their associated control systems are unable to distinguish between genuine consumption due to energy users or energy loss via leaks.

Another reason for energy loss in the distribution system is due to heat loss. Heat is lost from all pipes distributing hot fluids (e.g. steam, hot water or other heated contents) to the surrounding environment. The hotter the content in relation to the surrounding environment, the larger are the losses. Insulation is used to minimize heat losses. The thicker the insulation, the smaller is the heat loss. However, no amount of insulation will stop the heat loss from the distribution system altogether.

Modern day thinking about energy reduction

Similarly, if the distribution system is transporting cooled or chilled fluids (e.g. cooling water, chilled water or cooled content), heat is 'lost' from the area surrounding the distribution network – energy is gained by the distribution system. Similar to heat lost, no amount of insulation will stop the heat gain of the distribution system altogether.

In order to meet the demands of the energy user, the amount of energy loss (from heated pipes) or gain (from chilled pipes) will need to be compensated for by the energy generation system. These losses are known as distribution efficiency losses and can also be quantified by the energy efficiency equation.

Energy could also be lost by the energy user. Similar to the energy generation system and the energy distribution system, no energy user is 100 per cent efficient. A similar logic and energy efficiency formula can be applied to all of the energy uses.

The final way for energy losses in the distribution system to occur is a little counter-intuitive – examples of system losses have been mentioned in an earlier section of this chapter ('A machine can be energy efficient and still use more energy') when the organization installs the most efficient equipment but the system as a whole remains inefficient. For example, a chiller installed on a site could be the most efficient chiller available on the market. However, the design of the pipework distributing the chilled water in the plant could make the chilled water system inefficient. This could be caused by:

- throttling valves – there is a valve which is partly closed against the outlet of a pump – sometimes, the pump and valve could be very far away from the chiller, perhaps closer to the end user;
- pumps in series – there is a distribution pump supplying another distribution pump which in turn supplies another distribution pump, before getting to the end user;
- attemperation – there is a bypass line supplying chilled water from the outlet of the chiller back to the inlet of the chiller;
- a knock-on effect from other energy efficiency projects – energy efficiency projects have been implemented on other equipment that use chilled water and, as a result, the load on the chiller becomes smaller and outside the energy-efficient region of operation;
- an over-optimized machine – to tap into the high efficiency of the chiller, there is a specific design criteria or design requirement, which was not followed during the design or installation of the pipework.

Energy inefficiency in a system can also arise from a poor operating regime, poor maintenance and poorly executed modifications. These account for approximately 50 per cent of the cases in industry, commerce, and in a building. Figure 3.9 shows the root causes of energy inefficiencies that can be found.

Energy Audits

Chapter 3 Understanding energy use

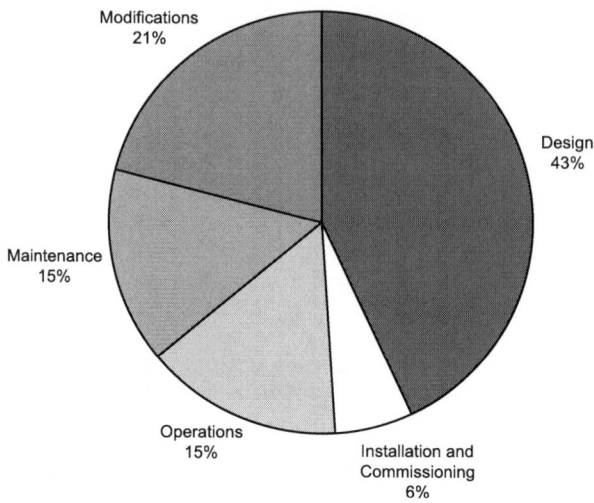

Figure 3.9: Root cause of energy inefficiencies

The same formula for energy efficiency can be applied to the whole generation, distribution and end use of energy by summing up all of the energy inputs and energy losses in a system before calculating the energy efficiency. This is known as 'system' efficiency.

When identifying energy reduction opportunities as a system, organizations need to first identify the end users of energy needs and then assess the efficiency as a 'whole, interdependent system': from the energy generation, through its distribution, to the said end user.

Figure 3.10 shows the relationship between energy purchasing, energy generation, energy distribution and energy use that needs to be taken into account. If the site generates waste, there may also be energy use in the waste segregation and waste treatment processes. Examples of other common energy use that can be assessed as a system are: a complete manufacturing process, a whole building, and steam, hot water, cooling water, chilled water, chilled glycol, compressed air, nitrogen, vacuum, heating, ventilation and air conditioning systems.

Using a systems approach also realizes additional benefits: reduced maintenance of the heat exchanger and additional time available for manufacturing, and increased sales as a result.

Modern day thinking about energy reduction

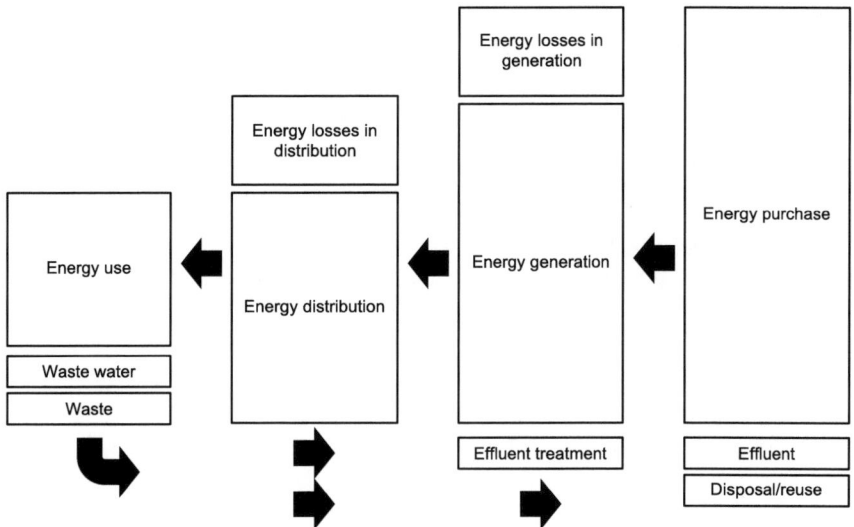

Figure 3.10: Relationship between energy purchase and end use

New build

Energy inefficiencies can also be located in newly installed and commissioned plants. These are mainly due to inefficient designs and/or poor commissioning and installation. Collectively, they account for the other 50 per cent of the energy inefficiencies encountered in industry and in commerce. (See Figure 3.9).

During the operational phase of the life cycle, many opportunities involving large capital costs are not economical. This is because the project has to pay back all of the allocated capital. In a new build, the organization spends the capital to gain the beneficial use of the new build. Energy efficiency is assessed on a margin cost basis, i.e. the additional cost required to obtain additional savings in energy.

As such, it is prudent that, when an organization is planning for major refurbishment, renovation, retrofitting and/or building new plants or buildings, the organization and its design team should assess the energy consumption and energy efficiency of the design and incorporate improvements right from the beginning. For companies utilizing an energy management system based on ISO 50001, the design assessment should consider the planned or expected lifetime of the equipment.

Incorporating energy efficiency features at the design stage gives the organization three benefits:

Chapter 3 Understanding energy use

1) the quantum of energy consumption can be minimized by matching the actual demands of energy with the supply;
2) it minimizes the capital cost to build; and
3) it minimizes the cost of subsequent energy efficiency retrofits.

In the context of a building, incorporating energy-efficient design principles in Genzyme Corporation's 12-story LEED Platinum headquarters in Cambridge, Massachusetts, uses 42 per cent less energy and 34 per cent less water, and boosts employee motivation and productivity by 15 per cent, to a comparable building.[64] Closer to home, the landmark London 2012 Olympic Games building, the Velodrome, is a good example of incorporating green building principles from the start.

The Velodrome achieves a 31 per cent better energy performance over Part L of the Building Regulations (2006) on energy efficiency. This building incorporated the following at the design stage:[65]

- natural ventilation for most areas of the building;
- low-power, active ventilation, utilized only for conference rooms and server rooms with short duct runs;
- the volume of air in the building minimized by utilizing a double curvature roof;
- glazing utilized to introduce daylight into the building and minimize artificial lights;
- exposed concrete in most of the building areas acting as thermal mass to modulate the temperature, with night-time purging to refresh the air;
- improved insulation of the building envelope;
- low-temperature underfloor heating and rapid response air heating used for events and functions;
- energy-efficient mechanical and electrical system utilized.

As stated before, incorporating energy efficiency at the design stage does not need to be limited to buildings. The same can be applied to industrial processes. Pfizer's County Cork plant in Ireland consumes approximately 6 million m^3 of nitrogen (N_2). Traditionally, Pfizer purchased all its high-purity N_2 from an external supplier. In 2008, the company decided to install an on-site N_2 generation system to displace about two-thirds of the N_2 consumed on-site. Energy-efficient design reviews identified seven opportunities for improvement on the proposed design, giving an energy reduction of 21 per cent on top of the operating cost savings.[66]

Resource efficiency

An organization focusing on managing its utilities efficiently and effectively is only tapping into a small fraction of the available savings. A

2degrees survey of more than 700 organizations globally found that they are beginning to look into reducing energy, water and waste simultaneously.[67]

As shown in some of the earlier examples, energy-saving projects can also give other benefits, such as raw material savings, waste reduction, improved quality, removal of equipment bottlenecks, improved yield, reduced water consumption and effluent disposal, reduced transport and logistics requirements and, sometimes, freeing up human resources for other tasks. This gives the maximum possible cost reduction: energy, water, waste and other resources.

There is another saying: 'one man's waste is another man's food!' This is also true in energy. The example of Coca Cola HBC (see Chapter 2) comes to mind. Another good example is Dawn Meats, a supplier of beef and lamb for food retail and food services markets throughout Europe. Dawn Meats consumes natural gas to generate steam. A typical steam boiler is 80 per cent efficient and 20 per cent of the energy is emitted into the environment via the stack (or chimney) in the form of 150 °C hot exhaust gas. Some of the steam is consumed to generate hot water for cleaning and sanitization purposes. By applying horizontal thinking across departments, Dawn Meats installed a state-of-the-art heat pipe heat exchanger to condense the hot exhaust gas to generate hot water. This recovered hot water is used to preheat water used for cleaning and sanitization and reduces natural gas consumption by 18 per cent.[68]

Sometimes, energy auditing on a systems basis only quantifies the energy-saving portion of the opportunity. This could be due to the energy auditor being unable, or not knowing enough about the organization, to quantify the other benefits. It could also be due to a pigeonholed mentality of managers to classify opportunities based on corporate initiatives or based on the department's budget. Both practices can lead to poor investment decisions where one department refuses to pay for the most cost-effective energy opportunity because they won't realize all of the savings themselves.

The pigeonholed mentality may also stem from the appearance of the word 'energy' in an energy audit. Some organizations are beginning to call this type of opportunity finding exercise 'resource efficiency audits'.

To fully appreciate and take advantage of resource efficiency-based opportunities, all sides of the savings should be quantified and utilized to calculate returns on investment. Many of these benefits can be quantified with the aid of the managers within the organization.

Supply chain efficiency

An even larger picture in which efficiency gains and energy reduction can be realized is provided by supply chain collaboration. Chapter 1 gave an example from Walkers Crisps, which realigned its purchasing specification to minimize its overall product carbon footprint.

To be able to tap into supply chain efficiency requires a thorough understanding of the operating parameters of an organization and how energy is consumed within the organization. With this in-depth information, and information about the energy consumption of each entity within the supply chain, the organization can facilitate a collaboration with each entity to identify opportunities to optimize the supply chain as a whole.

When Proctor & Gamble (P&G) began to assess its impact on the environment, it carried out a life cycle analysis of the whole supply chain. The findings showed that the largest energy consumption was from its laundry products by its customers, due to the need to heat water in the washing machine for washing. According to Len Sauers, Vice President of P&G's global sustainability initiative, 'This surprised me. I thought our highest energy impact was going to be somewhere in the manufacturing process.' P&G set off working on creating a product that enables its customers to wash in cold water without compromising the performance of the laundry detergent in hot wash cycles.[69]

P&G's competitor, Unilever, took the concept further and included water savings. An assessment of Unilever's laundry products showed that around 38 per cent of all domestic water is used to clean clothes. In water-scarce places, water is an expensive resource. Unilever's solution is called 'Comfort One Rinse' [70] – saving up to 50 per cent of domestic water consumption and the corresponding energy reduction.

Figure 3.11 shows the relationship between an organization's life cycle and its supply chain collaborations.

This category of opportunities may also include other organizations and entities not in the supply chain. This is especially prominent where one organization has excess heat that is sold or shared with neighbouring organizations or neighbouring households. An innovative £3.9 million partnership between Islington Council, UK Power Networks and Transport for London will see the waste heat from London Underground's Northern Line ventilation shaft and from UK Power Networks' high-voltage substation being put to good use. This project will connect these sources of waste heat, currently vented into the atmosphere, to an existing Bunhill Heat and Power plant, supplying heat and power to approximately 1,200 homes and a leisure centre. The addition of waste heat to the plant will reduce its CO_2 emissions by a further 500 tons of CO_2 per year.[71]

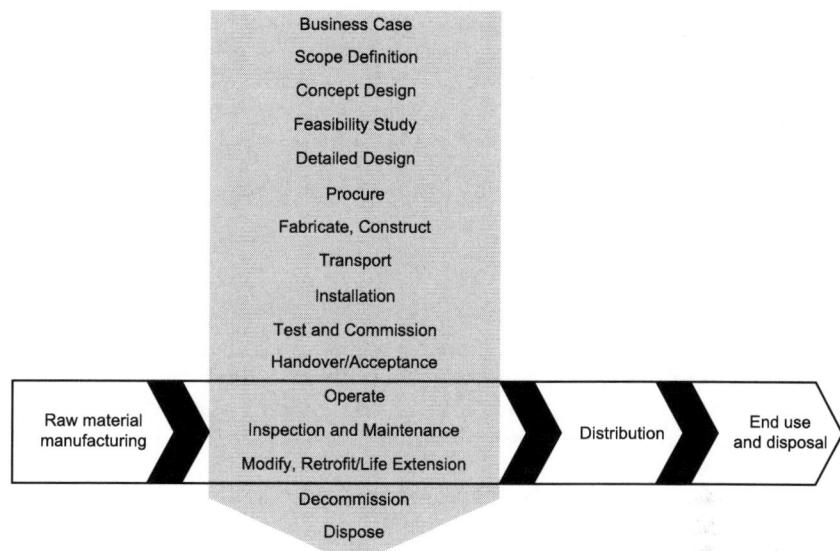

Figure 3.11: Relationship between business life cycle and supply chain

Chapter 4 Defining the boundaries of an energy audit

Chapter 2 described the various reasons why an organization may want to conduct an energy audit and the generic types of energy audits available in the marketplace. Chapter 3 builds on them by describing several ways in which an energy audit scope could be defined: by equipment, by systems and/or on a whole plant, whole building or whole organization basis. There is one additional piece of information necessary to completely define the boundaries of an energy audit – the minimum necessary to meet regulatory requirements: the Energy Efficiency Directive (EED).

Energy Efficiency Directive (EED)

Introduced on 25 October 2012, the EED is one of the key framework directives to improve energy efficiency. This directive amends, consolidates and modifies several EU efforts to improve energy efficiency, to achieve a European-wide 20 per cent primary energy reduction (or 368 million tons of oil equivalent (Mtoe)) by 2020.

It posits that energy efficiency will disseminate innovative technologies to rejuvenate aging and less efficient equipment, will reduce GHGs to mitigate climate change, will create new jobs and will increase the competitiveness of Europe.

The majority of the articles are revisions or a consolidation of previous directives that the EED replaces. There are two new requirements in the EED: energy management and energy audit responsibilities for four types of organizations:

1) central governments and other public bodies;
2) energy distributors and energy suppliers;
3) SMEs; and
4) large enterprises.

This book does not cover all of the requirements of ESOS. It describes aspects of ESOS pertinent to an energy audit only. Additional information and guidance about EED is available at the Directorate-General for Energy website. The EED defines an organization as an SME if it employs less than 250 people, has an annual turnover not exceeding €50 million and/or has an annual balance sheet not exceeding €43 million.

Once an organization has ascertained that it is categorized as non-SME, the EED requires the organization to carry out an energy audit by 5 December 2015 and every four years thereafter.[6] Even though the European Commission mandated the development of EN 16247, the EED does not, in fact, require companies to use EN 16247 standards when carrying out an energy audit. Organizations can choose one of the following three methods to use to comply with the EED:

1) energy audit carried out as part of a wider ISO 14001- or ISO 50001-based management system certified by a recognized certification body;
2) energy audit carried out by a qualified and/or accredited expert (these may be in-house experts when supervised by a relevant authority or externally sourced experts);
3) the output of the energy audit conforms to the minimum standards set by the EED, i.e. the energy audit:[72]
    - is 'based on up-to-date, measured, traceable operational data on energy consumption and…load profiles';
    - 'comprise[s] a detailed review of the energy consumption profile of buildings or groups of buildings, industrial operations or installations, including transportation';
    - (whenever possible) is based 'on life-cycle cost analysis (LCCA) instead of… [simple paybacks] in order to take account of long-term savings';
    - is 'proportionate, and sufficiently representative to permit the drawing of a reliable picture of overall energy performance and the reliable identification of the most significant opportunities for improvement';
    - 'shall allow detailed and validated calculations for the proposed measures so as to provide clear information on potential savings';
    - 'data used… shall be storable for historical analysis and tracking performance';
    - does not have clauses in the contract and report that prevent the findings from being transferred for implementation by a different service provider.

Applicability of the EED in the UK

In the UK, the requirement to carry out an energy audit is transposed into UK law as the Energy Savings Opportunity Scheme Regulations 2014

---

[6] The compliance date is 5th December 2015 and every 4 years thereafter, i.e. the date by which an EED-compliant energy audit is to be completed.

Chapter 4   Defining the boundaries of an energy audit

(ESOS). This book does not cover all of the requirements of ESOS. It describes aspects of ESOS pertinent to an energy audit only.

Additional information and guidance about ESOS is available at the Department of Energy and Climate Change's (DECC) website. (See www.gov.uk/energy-savings-opportunity-scheme-esos)

ESOS participants – Large undertakings

The UK government interprets 'enterprises' mentioned in EED as 'undertakings'. This is defined in paragraph 1161 of the Companies Act 2006 as a body corporate or partnership or an unincorporated association carrying on a trade or business with or without a view to profit. It includes limited companies, public companies, trusts, partnerships, unincorporated associations and not-for profits engaged in a trade or business.[73]

Large undertakings in the UK, either independently or as part of a corporate group with more than 250 people in the UK or having an annual turnover exceeding €50 million and having an annual balance sheet exceeding €43 million are required to participate in ESOS,

where:

- For large undertakings located in several EU member states, the participation of the organization is to be assessed within the boundaries of each member state and use their rules and regulations. Additional information is available from each member state and at the Directorate General for Energy website.
- Within the UK:
  o When determining the number of staff, including directors, this is to be computed as the average of the monthly totals through the ESOS qualifying period.
  o When converting from Euros to Sterling, the Bank of England Sterling to Euro spot rate on the prevailing qualification date is to be used. (See www.bankofengland.co.uk/boeapps/iadb/Rates.asp?Travel=NIxRSx&into=GBP)
  o When converting between energy units, the latest UK Government conversion factors for Company Reporting are to be used. (See www.ukconversionfactorscarbonsmart.co.uk/)
  o For corporate groups consisting of more than one group of companies in the UK, if one or more groups meets the definition of a large undertaking, the whole corporate group will be required to participate in ESOS, regardless of whether the parent company is in the UK. The undertakings may wish to aggregate or disaggregate in their participation in ESOS.

Applicability of the EED in the UK

- o Universities not subject to the Public Contracting Regulations 2006 in England, Wales and Northern Ireland and the Public Contracts (Scotland) Regulations 2012 in Scotland, are included in ESOS.
- o A collegiate-based university with separate central and college governing bodies is to assess their participation as separate enterprises.
- o For organizations using subcontractors to carry out part of their operations and do not directly pay for those energy costs involved, the subcontracted portion is excluded from ESOS.[7]
- o The eligibility of franchisee and franchisor organizations are assessed separately.

The UK government estimates that around 9,400 large enterprises are within the scope of ESOS. They operate 170,000 to 200,000 buildings, of which 8,000 to 10,000 are for industrial purposes. The UK government also estimates a total potential of 43 TWh (or 43,000,000,000 kWh) can be identified from the energy audit with a payback of two years or less and that 0.7 per cent of this potential will be realized. The financial benefit is worth £1.6 billion over a 15 year period for the large undertakings as a results of energy cost savings. The policy will also generate additional benefits from improved air quality and savings from avoided CO_2 emissions credits.[74]

ESOS assessment

ESOS requires large undertakings to carry out an ESOS assessment by 5 December 2015 and every four years thereafter. An ESOS assessment is:

- An assessment led by an ESOS lead assessor from an approved register of assessors held by the Environment Agency. The ESOS lead assessor is competent according to PAS 51215 and leads, reviews and approves the ESOS energy assessment. The ESOS lead assessor may also be part of the energy audit team when carrying out the energy audit.
- A computation of the large undertakings' total energy consumption, including energy use in the buildings, processes and transport that it consumes or pays for, excluding grey mileage – mileage accrued by employees travelling to and from a single place of work and which is not paid for by the organization. For transport energy consumption, all aviation and shipping journeys beginning or ending in the UK and all fuel consumption on roads and rails in the UK is included in ESOS.
- Using the same energy units in the above bullet point, determine the large undertaking's significant areas of energy use, i.e. the energy use that covers at least 90 per cent of its total energy consumption.

---
[7] The subcontractor has to assess its ESOS participation independently from that of the organization.

Energy Audits

Chapter 4   Defining the boundaries of an energy audit

This is the boundary where an energy audit is required. The remaining up to 10 per cent is known as *de minimis* energy consumption where an energy audit is not required.
- Energy consumption covered by either a certified ISO 50001 energy management system, Display Energy Certificate (DEC) with the accompanying recommendation report, and Green Deal assessment reports are deemed to be equivalent to an energy audit and can be excluded from the boundary.
- An energy audit is required for all remaining energy consumption in the boundary. The ESOS compliant energy audit may consists of one energy audit, several smaller energy audits and/or energy audits spread over four years.

The ESOS lead assessor signs off the ESOS assessment. The organization completes an ESOS evidence pack, has the ESOS results reviewed and signed off by a board-level director (or in the absence of a board, a senior manager) and submits the evidence pack to the Environment Agency before the compliance due date.

The Environment Agency is the overall scheme administrator for the United Kingdom. The Environment Agency, Scottish Environment Protection Agency, Natural Resources Wales and Northern Ireland Environment Agency will be regulators (ESOS refers to this role as the 'compliance body') for England, Scotland, Wales and Northern Ireland respectively. The Environment Agency has overall responsibility for ESOS as 'scheme administrator', and has the authority to apply penalties, enforcements and appeals, and 'name and shame' organizations that are not in compliance with the requirements of ESOS.

Implications of ESOS on energy audits

In terms of complying with the energy management and energy auditing requirements of the EED and ESOS, both the directive and ESOS does not specify which type of energy audit or what level of detail is necessary. In the UK, as long as the energy audits making up the ESOS assessment conform to the minimum requirements, it is compliant.

ISO 50001, ISO 50002 and EN 16247 on their own, do not fully conform to the requirements of the EED and ESOS. Organizations should fully understand the requirements specified in the EED and ESOS in order to be able to specify the required energy audit and to be fully compliant. Alternatively, organizations may choose to engage an ESOS lead assessor to assess their eligibility, and define the boundary and scope of the required energy audit. Table 4.1 compares the energy audit requirements of EED and contrasts it against ISO 50001:2011, clause 4.4.3 'Energy review', EN 16247-1, and ISO 50002.

Table 4.1: Comparison of energy audit requirements

Requirements		Standards		
EED	ESOS	ISO 50001	EN 16247-1	ISO 50002
'Up-to-date, measured' and 'traceable' energy data.	Based on a minimum of 12 months of verifiable and data drawn from the immediate 24 months before the start of the energy audit. If less than 12 months of data is used, this is to be explained in the ESOS evidence pack.	Up-to-date, measured energy and other data, which are traceable.	Historical data provided by the organization, or are measured by the energy auditor, and are traceable.	Historical data provided by the organization, or are measured by the energy auditor based on representative sampling, and are traceable. ISO 50002 has provisions for checking and verifying the accuracy and repeatability of data.
Detailed review of the whole organization.	Total boundary of assessment covers 90% of all energy consumption including buildings, processes and transport less areas already covered by a certified ISO 50001 energy management system, DEC or Green Deal assessment.	Bound by scope and boundaries of the energy management system.	Bound by scope and boundaries specified by the organization.	Bound by scope and boundaries specified by the organization.
Based on overall energy performance.	There is no specific requirements to assess	High-level assessment of significant energy users.	Rigorous assessment within the scope. Energy	Rigorous assessment within the scope. Energy

Chapter 4 Defining the boundaries of an energy audit

Requirements		Standards		
EED	ESOS	ISO 50001	EN 16247-1	ISO 50002
	the overall energy performance of the organization. Assessing the energy performance arising from opportunities for improvement is implied. A breakdown of energy consumption and its variation over time is required.	Energy performance indicators could be based on overall energy performance, system performance or equipment performance.	performance indicators could be based on the energy efficiency of equipment or of the system.	performance indicators could be based on the overall energy use, energy consumption and energy efficiency of equipment or of the system.
Identification of the most significant opportunities based on energy efficiency.	Identification of opportunities for energy efficiency improvement that are in the operational control and influence of the ESOS participant.	Identification of significant energy use and opportunities to improve the energy use within the scope and boundaries of the energy management system.	Identification of opportunities to improve within the scope, boundaries and investment criteria agreed with the organization.	Identification of opportunities to improve within the scope, boundaries and investment criteria agreed with the organization.
'Detailed and validated calculations'.	Use appropriate calculation to determine potential savings in terms of potential	None.	Based on detailed calculations. Calculations are not validated by an external party.	Assessment is based on the level of detail required by the organization.

Implications of ESOS on energy audits

Requirements			Standards		
EED	ESOS	ISO 50001	EN 16247-1		ISO 50002
	energy and monetary savings.				Calculations are not validated by an external party.
	ESOS lead assessor to review, approve and sign off the ESOS assessment which may include the energy audit.				
	ESOS evidence pack is presented in boardroom and sign off by a board director.				
If possible, based on life cycle analysis.	If possible and practicable, based on life cycle analysis.	Considers planned life during purchasing.	Organization specifies the method to quantify opportunities.		Organization specifies the method to quantify opportunities. When evaluating financial viability, ISO 50002:2014, Clause 5.7.4 recommends that the 'planned or expected operating lifetime' is considered.
No clauses to prevent the findings from	None.	None.	None.		None.

Energy Audits

Chapter 4 Defining the boundaries of an energy audit

Requirements			Standards		
EED	ESOS	ISO 50001	EN 16247-1	ISO 50002
being transferred to another qualified/accredited energy service provider.					

On applying an energy management system

When an organization defines the scope and boundaries of its energy management system, the easiest way to comply with ESOS is to define it in a manner that encompasses the full scope of the organization. This allows the energy review to capture the whole organization, the energy performance of the whole organization to be assessed, and the most significant opportunities to save energy to be identified. If the ISO 50001-based energy management system does not cover at least 90 per cent of either the organization's total energy consumption or the organization's total energy cost, it is required to cover the remainder via an energy audit.

The ISO 50001 energy review tends to be a high-level overview rather than a document with detailed calculations and using life cycle analysis for every identified opportunity. An organization needs to take a view that the level of detail required and the method used to evaluate return on investment be commensurate with the cost involved, speed to implement, and level of business risk. A type 1 energy audit could also be used to prepare the ISO 50001 energy review.

The energy review may contain opportunities that can be implemented using little or no capital cost. Some examples are housekeeping, better scheduling, insulation, simple control systems, or operating a more efficient machine for the given duty. A return on investment criteria such as simple payback or a discounted cash flow analysis may be sufficient for these opportunities. Carrying out a detailed assessment and life cycle analysis for the low-cost and no-cost opportunities significantly increases the cost and renders the opportunity uneconomical.

For opportunities that involve high capital costs, have long lead times to implement or have a high business risk, a more in-depth study with detailed calculations and a financial evaluation such as a life cycle analysis would be in order. A type 2 or type 3 energy audit comes in handy at this stage.

One key benefit of ISO 50001, as suggested in Chapter 1, is that energy savings can come from removing an energy user, reducing the quantity of energy consumption or by improving efficiency. Using an office lighting example, energy savings can come from:

- Energy use: the use of lighting in this area is not necessary as light from the corridor also lights this area.
- Energy consumption: the hours in which the lighting in this area can be controlled in line with demand using a passive infrared detector (PIR).
- Energy efficiency: the light fittings can be replaced with LEDs which gives the same light output with significantly reduced electrical input.

Chapter 4   Defining the boundaries of an energy audit

ISO 50001 opens up the number of opportunities organizations can use to achieve savings. Another key benefit of ISO 50001 is the buy-in and active participation of senior management. When using ISO 50001, senior management is required to actively participate in setting the energy policy, putting the relevant resources in place and making sure improvements occur as planned.

On applying an environmental and/or sustainability management system

Similar to an energy management system, when an organization defines the scope and boundaries of its environmental and/or sustainability management system, it has to encompass the full scope of the organization. This allows the environmental aspects of the whole organization to be captured, the energy performance of the whole organization to be assessed, and the most significant opportunities to save energy to be identified.

From an environmental perspective, the energy performance is measured and reported through CO_2 emissions. While an organization with a certified ISO 50001 energy management system is exempt from ESOS, an organization with a certified environmental and/or sustainability management system is not exempt from carrying out an energy audit.

Alternatively, if the organization chooses to, it would be well placed to develop and integrate the requirements of ISO 50001 into its existing environmental and/or sustainability management system. The new system could be certified by a certification body and the organization would be exempted from carrying out an ESOS assessment.

On energy auditing

The EED and ESOS do not specify which methods to use for an energy audit, nor what level of detail must be contained in an energy audit. An organization is free to choose which methods are to be used, e.g. ISO 50002, DECs and accompanying advisory reports, Green Deal assessments or any other methods.[8]

As mentioned earlier, managers need to remember that the total scope of the energy audit being commissioned must be sufficiently broad to cover 90 per cent of either the organization's total energy consumption or the organization's total energy cost, and that the energy audit use at least 12 consecutive months of energy and other related information from the immediate 24 months prior to the energy audit commencing.

---

[8]   Many experts question the equivalence of an energy audit carried out to ISO 50002 to that of a DEC or a Green Deal assessment. However, this has been included in ESOS.

When commissioning energy audits, the organization can choose to have one energy audit to cover the full scope of the organization, or choose to commission several energy audits to cover different aspects of the organization and pull the results of the energy audits together, to form the basis of an ESOS evidence pack.

By using this information and having a defined scope and boundaries of the energy audit(s), a reliable picture of the organization's energy consumption, its relationship to factors that cause it to vary and opportunities for improvement can be identified.

If the organization has carried out several energy audits within the four-year compliance period but they do not cover at least 90 per cent of either its total energy consumption or its total energy cost, it is required to carry out an energy audit for the remainder.

In the energy audit, a breakdown of energy consumption within the organization is required. This could take the form of a pie chart or any other form of presentation and aggregation. Organizations may find the breakdown useful in prioritizing an energy audit. A series of breakdowns can also be compared and trended to show the changes in energy use and consumption. This could also be another source of identifying inefficiencies.

The energy audit also needs to identify cost-effective improvement opportunities that are within the operational control of the organization. The criteria for 'cost-effective' and what constitutes 'within the operational control' needs to be defined prior to the start of an energy audit. For example, if an energy audit is commissioned by a building tenant, opportunities for improvement should relate to those the tenant can implement. Opportunities related to the control and responsibility of the building owner are out of scope. When using ISO 50002, these items will be covered in the planning phase of the energy audit.

Scoping the energy audit

It follows that, when scoping the boundaries of an energy audit, one of the very first tasks for an organization is to determine if it qualifies as a large undertaking. This is achieved by assessing the headcount and assessing the details located in the organization's accounts: balance sheets, profit and loss statements, and cash flow statements.

The next step is to compile and list the energy consumption and/or energy costs of the organization. According to the UK Government, between 4,400 and 6,400 organizations required to participate in ESOS participates in environmental permitting regulation, Climate Change Agreement (CCA), EU Emissions Trading Scheme (EU ETS) and CRC. The organization may already have this information, especially from its

Chapter 4   Defining the boundaries of an energy audit

manufacturing processes and buildings, through to its environmental permits reports, CCA reports, EU ETS reports, CRC reports, environmental reports, corporate social responsibility (CSR) reports and sustainability reports.

One area where an organization may get caught out is when computing its energy consumption or costs from transport. ESOS includes the energy costs from transport for which the organization pays. This includes the petrol, diesel and electricity the company pays for when its employees go on business trips, for the transportation of goods from the factory to the customer, and for fuel costs on company cars. For air and sea transport, the fuel consumption of journeys beginning or ending in the UK is included in ESOS. For road and rail transport, the fuel consumption while travelling in the UK is included in ESOS.

Large organizations with many buildings may choose to list the energy consumption and energy costs by building, process and transport. If the data are available, the organization may also choose to list the energy consumption by energy use or by energy type. Other organizations may want to group them in a different manner, e.g. by geography or by business unit, in order to facilitate easier accounting and the selection of areas for energy auditing. Let us take a portfolio of five buildings, as shown in Table 4.2, as an example. Based on energy consumption, Building 5 will be prioritized for an energy audit, followed by Building 1, Building 2, Building 4 and Building 3.

Table 4.2: Selection of buildings based on energy consumption

Building identification	Electricity consumption (kWh/y)	Natural gas consumption (kWh/y)	Total (kWh/y)
Building 5	70,890,120	95,621,569	166,511,689
Building 1	68,809,133	85,261,659	154,070,792
Building 2	33,740,496	76,174,645	109,915,141
Building 4	19,211,961	26,149,858	45,361,819
Building 3	6,082,235	13,079,434	19,161,669

Then, the next task would be to simply rank them and choose to carry out an energy audit on the highest 90 per cent of the energy consumption.

If the organization's energy consumption is purely from a portfolio of buildings, it is possible to divide the energy consumption with the total floor area of the building to give energy consumption per unit floor

Scoping the energy audit

area. These figures can be compared with the 'best in class' or building energy consumption benchmarks. The difference between the actual and the benchmarked energy consumption gives the potential energy savings that can be achieved. If this information is available, the organization can make use of it by selecting buildings with bigger energy-saving potential as part of the scoping process.

Using the same portfolio of five buildings as an example, by computing their energy consumption per unit floor area (Table 4.3) and then comparing that with the 'best in class' performance of 292 kW/m^2/y for electricity and 440 kW/m^2/y for natural gas, we can identify buildings with significant opportunities for improvement. While Building 5 has the highest energy consumption, in terms of both electricity and natural gas (see Table 4.2), it is also the most energy-efficient building of the portfolio. For electricity, the organization should prioritize Building 2, Building 1, Building 3 and Building 4 for energy audits. For natural gas, the energy audit should prioritize Building 2, Building 3, Building 1 and Building 4.

Table 4.3: Selection of buildings based on energy consumption

Building identification	Electricity consumption (kW/m^2/y)	Natural gas consumption (kW/m^2/y)
Building 2	531	1,199
Building 1	432	535
Building 3	382	822
Building 4	309	420
Building 5	278	375

Similarly, if the organization consists of production processes and/or involves a large proportion of transportation, similar energy performance indicators can be developed and benchmarked to arrive at processes and transport operations that have bigger energy-saving potential.

The next task would be to discount from the scope the energy users that are covered by a certified ISO 50001-based energy management system, DEC assessments and Green Deal assessment. This gives the remaining scope for which an energy audit is required. Lastly, armed with the above information, the organization should choose a preferred method for energy auditing, and assess whether the energy audit should use internal resources, external resources or a combination of both. For organizations wanting to comply with the requirements of ESOS, energy audits should be led by a qualified lead energy assessor from an Environment Agency approved register. Qualified lead energy assessors are members of

Chapter 4   Defining the boundaries of an energy audit

professional institutions and associations where their members have been assessed to be competent according to PAS 51215, which is described in Chapter 7.

Depending on the energy maturity of the organization, it then chooses a suitable type of energy audit to cover the scope of the energy audit. If the organization is starting out on its energy management efforts and does not know where significant energy savings can be made, a type 1 energy audit covering the whole scope of the organization is a good place to start. It helps an organization to identify and zoom in on areas where significant opportunities can be achieved. Tables 4.4a and 4.4b show the energy use breakdown from 3 building sub-sectors and 16 industrial process sub-sectors. They show the major energy use in the sub-sectors and help to prioritize energy audit works. The organization can then carry out targeted energy audits on those energy users, using a type 2 or type 3 energy audit.

Table 4.4a: Energy use breakdown in three building sub-sectors

	All buildings	Commercial	Residential
Space Heating	39.6%	29.9%	46.4%
Space Cooling	10.3%	11.3%	9.6%
Lighting	9.9%	15.3%	6.1%
Water Heating	13.1%	7.5%	17.1%
Refrigeration	4.4%	5%	4%
Electronics	4.2%	3.3%	4.8%
Ventilation	2.8%	6.9%	0%
Computers	2%	2.7%	1.5%
Cooking	3.3%	2.6%	3.8%
Other	10.3%	15.4%	6.7%

(Source: Building Energy Data Book March 2012)

Scoping the energy audit

Table 4.4b: Energy use breakdown in 16 industrial process sub-sectors

	Process Heating	Process Cooling & Refrigeration	Other Process Use
Transportation Equipment	24.8%	3.4%	5.7%
Textiles	37%	3.7%	4.6%
Plastics & Rubber Products	35.1%	7.6%	3.6%
Petroleum Refining	80.6%	1.3%	4.9%
Machinery	25.9%	1.4%	3.5%
Iron & Steel	75%	0.4%	3%
Glass & Glass Products	82.1%	0.5%	0%
Foundries	64.2%	1.1%	3.2%
Forest Products	55.6%	2%	4.1%
Food & Beverage	51.2%	11%	4.5%
Fabricated Metals	47.8%	1.4%	2.4%
Electric & Electronics	20.3%	5.3%	3.9%
Chemicals	55.8%	3.6%	5.5%
Cement	87.7%	0%	0%
Alumina & Aluminium	38%	0.5%	1.4%
Manufacturing	56.3%	3.8%	5.7%

Energy Audits

Chapter 4 Defining the boundaries of an energy audit

	Electro-chemical Uses	Pumps	Fans	Compressed Air	Material Handling
Transportation Equipment	1.1%	4.2%	3.1%	3.4%	2.3%
Textiles	0%	6.5%	4.6%	4.6%	3.7%
Plastics & Rubber Products	0.4%	5.6%	4%	4.4%	3.2%
Petroleum Refining	0%	6.6%	1.1%	1.7%	0.3%
Machinery	0%	4.9%	3.5%	3.5%	2.8%
Iron & Steel	1.1%	0.9%	1.6%	1.5%	4.9%
Glass & Glass Products	0%	2%	1.5%	1.5%	1%
Foundries	0%	1.1%	2.1%	2.1%	5.3%
Forest Products	0.1%	8.4%	5.3%	1.2%	2%
Food & Beverage	0.1%	3.7%	1.7%	1.7%	1.4%
Fabricated Metals	1.4%	4.1%	2.7%	3.1%	2.1%
Electric & Electronics	1.9%	3.9%	2.9%	2.9%	1.9%
Chemicals	5.4%	5.9%	2.7%	6.2%	0.3%
Cement	0%	0.8%	1.2%	0.8%	0.8%
Alumina & Aluminium	42.3%	0.9%	1.9%	1.4%	5.2%
Manufacturing	2.9%	5.9%	3.2%	2.6%	2.9%

Scoping the energy audit

	Material Processing	Other Systems	Facility HVAC	Facility Lighting
Transportation Equipment	6.9%	0.8%	30.9%	8%
Textiles	10.2%	1.9%	16.7%	5.6%
Plastics & Rubber Products	9.6%	1.2%	17.1%	5.2%
Petroleum Refining	1.3%	0.2%	1.3%	0.2%
Machinery	7.7%	0.7%	34.3%	7%
Iron & Steel	1.3%	0.2%	7.7%	0.8%
Glass & Glass Products	3.1%	0.5%	6.1%	1.5%
Foundries	1.1%	0%	14.7%	2.1%
Forest Products	5.7%	2.8%	7.7%	1%
Food & Beverage	5.7%	1.5%	10.3%	2.4%
Fabricated Metals	6.5%	1%	18.9%	5.2%
Electric & Electronics	6.3%	1%	36.2%	7.7%
Chemicals	5.3%	0.4%	5.7%	0.8%
Cement	6.2%	0%	0.8%	0.4%
Alumina & Aluminium	1.4%	0%	3.8%	0.9%
Manufacturing	0%	1%	10.4%	2%

Chapter 4 Defining the boundaries of an energy audit

	Other Facility Support	Onsite Transport	Other Non-process
Transportation Equipment	3.1%	1.5%	0.8%
Textiles	0.9%	0%	0%
Plastics & Rubber Products	1.6%	0.8%	0.8%
Petroleum Refining	0.1%	0.1%	0.5%
Machinery	2.8%	1.4%	0.7%
Iron & Steel	0.4%	0.4%	0.5%
Glass & Glass Products	0%	0%	0%
Foundries	2.1%	1.1%	0%
Forest Products	0.3%	1%	2.7%
Food & Beverage	2.3%	1%	1.7%
Fabricated Metals	1.7%	1.7%	0%
Electric & Electronics	4.3%	0.5%	1%
Chemicals	0.7%	0.2%	1.5%
Cement	0%	1.2%	0%
Alumina & Aluminium	0.9%	0.9%	0.5%
Manufacturing	1%	0.7%	1.5%

(Source: Manufacturing Energy and Carbon Footprint, 2010 Data, US Department of Energy, Advanced Manufacturing Office)

Scoping the energy audit

If, on the other hand, the organization is mature in its energy management practices and/or knows where and how significant energy savings can be achieved, the organization can commission type 2 or type 3 energy audits specific to those energy users. Common opportunities for improvement for various energy use can be found in Chapter 6.

The Royal Bank of Scotland Group (the RBS Group), one of the banking giants in the UK, operates from a property portfolio of more than 2,500 buildings. In 2012, the RBS Group set a goal to reduce its energy consumption by 15 per cent by 2014 using 2011 as its baseline. The specific method chosen by the RBS Energy Team was the lean method, based on the principles of 'prepare, diagnose, design, implement and sustain'.[75]

Approximately 40 per cent of its energy consumption comes from retail banking and data centres. The remaining 60 per cent comes from 600 multi-occupancy buildings (MOPs). MOPs house RBS back office functions and investment trading functions – although the occupants in MOPs are essential parts of RBS operations, they are less critical than data centres. MOPs are also buildings where the energy team has a large influence over their energy consumption.

Half-hourly electricity data and natural gas data were obtained and analysed according to a predetermined matrix. This matrix consisted of annual electricity consumption, annual natural gas consumption and energy consumption versus operating hours, degree-days and occupancy. The RBS Group identified 40 MOPs where their energy consumption did not match any known patterns of consumption. These 40 MOPs were prioritized for detailed investigation.

As an alternative to the whole building or whole process energy audits described above, the organization may also wish to define its energy audit boundaries and scope, and source its energy auditors, in a different manner. For example, the organization may want to appoint a specific energy audit for a specific area, say resource A for all ventilation systems and resource B for all heating systems, and use internal resources to carry out the energy audits on its own manufacturing processes. The organization may also choose to utilize a rolling programme of energy audits, covering all business areas over a fixed period of time.

As far as ESOS is concerned, as long as the energy audit's total boundaries cover 90 per cent of the organization's total energy consumption or total energy cost, and it meets the requirements of ESOS and PAS 51215, it is deemed to be compliant.

ArcelorMittal Saldanha Works in South Africa is a good example of implementing an ISO 50001-based energy management system and energy audits. ArcelorMittal Saldanha Works, located on the west coast of South Africa, manufactures Hot Rolled Coil steel products for the

Chapter 4 Defining the boundaries of an energy audit

sub-Saharan markets. Energy costs are approximately 44 per cent of the site's total operating costs.[76] In 2008, due to rising energy prices, rolling blackouts and the global economic downturn, ArcelorMittal needed to significantly reduce its energy costs in order to compete with their competitors.

Prior to implementing an energy management system and systems-based energy audits, the site implemented many energy reduction projects, many of which have been implemented on an ad hoc and departmental silo basis. In 2012, the site formed an energy team to implement a management infrastructure (ArcelorMittal's energy management system) covering three of the most significant energy uses, accounting for 70 per cent of the site's energy consumption. The site's senior management allocates 25 per cent of the site's budget towards improving its energy performance.

According to Reinet van Zyl, ArcelorMittal's energy manager

> The risk, with the focus on energy, was that it could have been just another initiative and as a new priority would have come along the energy management system implementation could have been dropped as a result of too few people being dedicated to the cause. ... Implementing an energy management system is the only way to ensure that the knowledge and practices are captured and institutionalised within the corporate culture and not reliant on any specific individual.[77]

Using a systems-based energy audit approach, the energy team identified 12 energy reduction projects using staff awareness, implementing opportunities not requiring any capital cost and without any impact on production and productivity levels.

Within the same year, the site saved 80 GWh/year or 5.3 per cent of its energy consumption, including a 26 per cent reduction in LPG. The value of the savings is calculated to be worth R90 million with a capital cost of R0.5 million. These projects have a payback of four days. By April 2013, an additional three projects were completed, giving a total energy savings of 6.6 per cent and a reduction of its specific energy consumption from 25.1 to 23.7 GJ/ton.

Why large enterprises only? Does it mean SMEs will not achieve any benefits?

ESOS and the examples in this book focus primarily on large enterprises. However, SMEs form 99.8 per cent of total UK enterprises.[78] Although the EED does not specify its applicability to SMEs, it requires each member state to put in place schemes to encourage SMEs to apply good management systems and to conduct energy audits as a basis to improve their energy performance.

Applying a management system for managing energy and for carrying out an energy audit also benefits SMEs. Der Grundhofer Vollkornbäckerei Peter Thaysen (Bäcker Thaysen) is an SME deploying an energy management system and energy auditing to manage and reduce its consumption. Based in Schleswig-Holstein, Germany, Bäcker Thaysen is a 950 m² craftsman bakery producing bread, buns, rolls, cakes and pastries. In the bakery, 80 per cent of its energy consumption is in the form of 180 °C to 360 °C thermal energy for baking. Analysis of energy data from the past four years showed that the bakery's energy mix is met by 28 per cent of electricity and 72 per cent of furnace oil per year. Both the electricity and the furnace oil cost the bakery DM68,300 (Deutschmark, pre-Euro figures).[79] The bakery uses its ISO certification as a marketing tool to differentiate its products from that of its competitors.

In the bakery, energy management is the responsibility of the production manager, who also has direct responsibility and authority for production scheduling and other production resources. The production manager used energy auditing as a tool to provide information about energy use and energy consumption within the bakery, and also to identify, analyse and quantify opportunities to reduce energy. During the energy audit, all production areas were analysed and an energy balance was produced, depicting the energy purchased, and its distribution and consumption by the energy users.

A significant majority of the recommendations covered the tightening of operational controls, installing local switches to enable machines and amenities to be turned off, and housekeeping and employee awareness issues. These were incorporated into the operational procedures and routines. Other opportunities centred on the insulation of hot and cold pipes in the distribution network, and carrying out regular calibrations of thermostats and thermocouples to minimize overheating and over-cooling.

The implementation of the recommendations from the energy audit resulted in a 6 per cent reduction in energy consumption; the energy consumption per kilogram of bakery product was reduced from 1.36 kWh/kg to 1.28 kWh/kg with a financial benefit of DM4,000 per year.

Reporting: Creating transparency for energy initiatives

In the UK, there are two new mechanisms requiring companies to report their efforts to improve energy efficiency: (1) the reporting of energy

Defining the boundaries of an energy audit

ngs under ESOS and (2) the mandatory GHG emissions
The Environment Permitting regulation, CRC, Climate Change
, and EU ETS also have energy reporting requirements. These
are not described in this book as they are existing reporting mechanisms and there are many materials available in print media and on the internet.

ESOS notification

Under ESOS, large organizations are required to maintain an ESOS evidence pack of records to substantiate the ESOS assessments, and their findings and recommendations.

The organization's board-level director (or senior manager in the absence of a board) is required to review, approve and sign off the ESOS evidence pack. This signifies that senior management understands the requirements of ESOS, and reviews the outcome of any energy audits including those in relation to ISO 50001 conformity and opportunities for improvement. If the organization utilizes an external lead energy auditor, one director or senior manager sign-off is sufficient. If the organization utilizes an internal lead energy auditor, two director or senior management sign-offs are required.

Once the sign-off has been obtained, on or before each compliance date, i.e. 5th December 2015 and every four years thereafter, the organization is required to notify the regulator of its compliance.[9] It is envisaged that this will be an online notification system where the organization will be required to answer a series of questions. The details of this evidence pack and reporting mechanism will be made available by the Environment Agency well before 5th December 2015, the first compliance date. DECC's ESOS guidance indicates that the following information may be required in the ESOS evidence pack:[80]

- Details of the organizations participating in ESOS.
- Details of the board-level director or senior manager.
- Details of ESOS lead assessor.
- Details of the organization's total energy consumption and areas of significant energy consumption.
- Details of the compliance method chosen.
- Any reasons for using less than 12 months of or unverifiable energy or other data in qualifying for ESOS participation and in the energy audit.
- Details of the energy audit methodology chosen.
- Any reason for not including an energy consumption profile in the energy audit.

---

[9] At the time of writing, the ESOS notification and reporting system is not operational. It is estimated that this will be fully operational before 5th December 2015.

Reporting: Creating transparency for energy initiatives

Mandatory greenhouse gas emissions reporting

A survey of business leaders by the Economist Intelligence Unit [81] agreed that sustainability issues, such as energy consumption and energy efficiency, are not discussed in the boardroom. Sixty-one per cent of the business leaders surveyed say sustainability initiatives benefit the organization. Yet, less than one third include any sustainability metrics in their annual reports and even less (20 per cent) assess the implications of sustainability on their business strategies.

The mandatory GHG emissions reporting regulation amends the Companies Act 2006 via the Companies Act 2006 (Strategic Report and Directors' Report) Regulations 2013, and requires all organizations listed on the main market of the London Stock Exchange to report their GHG emissions (CO_2, NO_x, methane and many other gases) in their annual reports. If the annual report does not contain this information, then it must point out the omissions.

Recently the European Commission too has approved and adopted a similar integrated reporting directive, to increase transparency and reporting of companies' responsibilities across the whole of Europe and on a wider sustainability and community basis.[82] This amends the Fourth (78/66/EEC) and Seventh (83/349/EEC) Company Law Directives and requires large organizations to disclose non-financial and diversity information in their annual reports in a uniform, EU-wide manner.

The purpose of reporting non-financial performance is to create visibility of sustainability and diversity, in order to put in place effective management and governance of these issues in the boardroom.

Instead of creating a new reporting framework, the UK and EU reporting requirement uses available non-financial reporting frameworks. This has its roots in Southwest Airlines. As one of the pioneers to develop and publish an integrated financial and non-financial report, Southwest Airlines reports its financial and sustainability performance in one unified report called *Southwest Airlines One Report*™. The report contains three sections: 'Performance' details the financial performance of the company, 'People' details customer satisfaction, and employee and community engagements, and 'Planet' details the energy and environmental performance of the company.

Southwest Airlines publishes the *Southwest Airlines One Report*™ online and posts a four-page summary to its shareholders. The report is well received by its shareholders, financial analysts and even its own employees and customers. An employee survey demonstrated a positive relationship between the organization's employees via the report and their job satisfaction. Its simple layout also allowed Southwest Airlines employees to answer customer questions more easily and direct them to additional information on the website.[83]

Chapter 4   Defining the boundaries of an energy audit

Since Southwest Airline's *Southwest Airlines One Report*™, there has been a proliferation of integrated reporting frameworks, each with varying degrees of success. Some of the most commonly used reporting frameworks are: the Global Reporting Initiative (GRI) Principles, the International Integrated Reporting Framework Principles, the Sustainability Accounting Standards Board (SASB) Standards, Arista 3.0®'s 'The Eleven Commitments', Arista 3.0®'s nine 'Integrity Principles', the United Nations-sponsored Principles for Responsible Investment (PRI) Initiative's principles, Natural Step's Four Principles of Sustainability, the UN Global Compact's 'The Ten Principles', the Earth Charter principles, and Corporation 20/20's 'New Principles for Corporate Design'.

These non-financial reporting frameworks include all items from the traditional annual reports (strategy and analysis, organizational profile and economic performance) and a host of non-financial aspects containing a range of elements, such as governance, commitment and engagement, environmental stewardship (materials, energy, water, biodiversity, emissions, effluents and waste, and compliance and transport), human rights (investment and procurement practices, non-discrimination, freedom of association and collective bargaining, child labour, forced and compulsory labour, security practices and indigenous rights), and labour practices and decent work (employment, labour–management relations, occupational health and safety, training and education, and diversity and equal opportunity).

Wide adoption of non-financial reporting breaks down the information asymmetry between annual reports and allows an organization's current state of play to be deciphered. Sustainability ratings such as the Carbon Disclosure Project (CDP), the Dow Jones Sustainability Indices (DJSI), the FTSE4Good Index Series and the Newsweek Green Rankings use this information to create organizational sustainability rankings – which are a potential source for competitive advantage.

As suggested in the Introduction and Chapter 1, the general public increasingly wants to consume products and services from environmentally friendly and sustainable organizations. The younger generations, too, look towards green credentials as one of the key employment criteria. These are two key factors that determine the longevity and long-term success of an organization.

Since the financial crash in 2008, there has been a rise in socially responsible investors who look beyond the strength of organizations' finance and into organizations' long-term strategies, the ability to execute their strategies and governance systems to track and deliver the strategies.[84]

Organizations that actively disclose their non-financial performance, raise their sustainability rankings and show a continual improvement in their ratings make themselves attractive to investors and build up the

Reporting: Creating transparency for energy initiatives

organization's balance sheet – another vital competitive advantage for organizations. Organizations that do not look towards longer-term sustainability may also face the risk of a shareholder and/or boardroom revolt or revolution – a phenomenon that is increasing becoming common since the economic crisis of 2008.

As such, the real value of mandatory information disclosure and sustainability ratings is not in the reporting or ratings, it is that they are benchmarks that focus organizations' attention and resources on challenging workplace practices and bringing about improvements. Two studies in the USA attest that organizations do take action when they are required to disclose information and are subject to being rated.[85]

Organizations working on supply chain initiatives also need information to assess their supply chain's energy consumption (or indirectly as CO_2 emissions). As indicated in Chapter 2, Tesco has a corporate aspiration to reduce the carbon footprint of the products it sells by 30 per cent; the availability of non-financial reporting helps to provide the information needed. As another example, in 2010, Walmart set an aggressive goal to eliminate 20 million tons of GHG emissions from its supply chain in 5 years.[86] To achieve this goal, Walmart asks its suppliers to answer and provide data, on an annual basis, to 15 environmental impact questions, and feeds these into a Sustainability Index and scorecards. Walmart's 'merchants' – the high-level managers responsible for the multibillion-dollar buyers who determine which products appear on its shelves – use the Sustainability Index as part of their purchasing decisions. As part of Walmart's merchants' performance review process, sustainability is one of the items reviewed and plays a role in determining pay rises and potential future promotions.[87]

Apart from supply chain initiatives, there may also be regional initiatives where a group of local organizations form local charters and partnerships to reduce energy consumption and improve sustainability performance. These local initiatives also need good non-financial reporting and ratings to assess the results on a like-for-like basis. One such regional partnership is the Low Carbon Oxford initiative.

Initiated by Oxford Strategic Partnership and Oxford City Council, 'Low Carbon Oxford (LCO) [88] is a pioneering city-wide programme of collaboration between private, public and non-profit organizations with the aim of ensuring Oxford's future as a sustainable and low carbon city.' The programme was launched on 14 October 2010 with 15 local organizations, including Oxford City Council, University of Oxford, Oxford Brookes University, Oxfordshire County Council, Mini Plant Oxford, Unipart, Stagecoach, Marks & Spencer and B&Q, supporting it.

At the heart of the initiative is a charter where each organization commits to, among others, organizational goals and targets to:[89]

- collaborate to reduce its carbon footprint across its sites and operations in Oxford by an average minimum of 3 per cent year on year for the next 10 years, giving a total of an 80 per cent reduction by 2020;
- reduce the organization's carbon intensity, if, during the reporting period, the carbon footprint increases due to growth in the number of sites, operations and output;
- issue annual progress reports on its targets.

Low Carbon Oxford uses the information provided by each organization and publishes the progress of the initiative as a group. Non-financial reporting using a common reporting framework is essential and forms an integral part of the initiative.

Chapter 5   The processes of an energy audit

Chapter 2 covered several types of energy audits that are available in the market and organizations can choose the different types of service to meet their needs. Chapter 4 provided a framework to define the minimum scope of an energy audit to comply with ESOS. Regardless of the energy audit type chosen and the breadth and depth of the energy audit, there are several generic processes in an energy audit. This chapter describes these steps when carrying out an energy audit.

EN 16247 and ISO 50002 have a defined energy audit process (see Figure 5.1) and defined output. This process is traceable, therefore it is called an energy audit. In contrast, ESOS is called an assessment because it only has a defined outcome or output. It does not have a defined procedure or process for generating the outcome.

In an energy audit, apart from defining the criteria for evaluating and prioritizing the energy audit findings, and final reporting, all other processes do not need to be performed sequentially. Some iteration of the processes may be necessary and required to arrive at a good and reliable conclusion, and to support the organization in making informed decisions.

Some reasons for the iterations may be: some associated equipment had originally been left out of the scope but is heavily interrelated or interdependent, data collected or measurements taken do not make sense, needing additional work, or the iterations could also be at the request of the organization. A good energy auditor may be able to estimate or include some iterations and/or rework. Excessive iterations and rework will incur additional costs.

Starting an energy audit without a clear definition of the objectives, and the scope and boundaries, could also lead to significant iterations, generate outputs with little benefit and ultimately lead to disappointment by the organization and the energy auditor.

processes of an energy audit

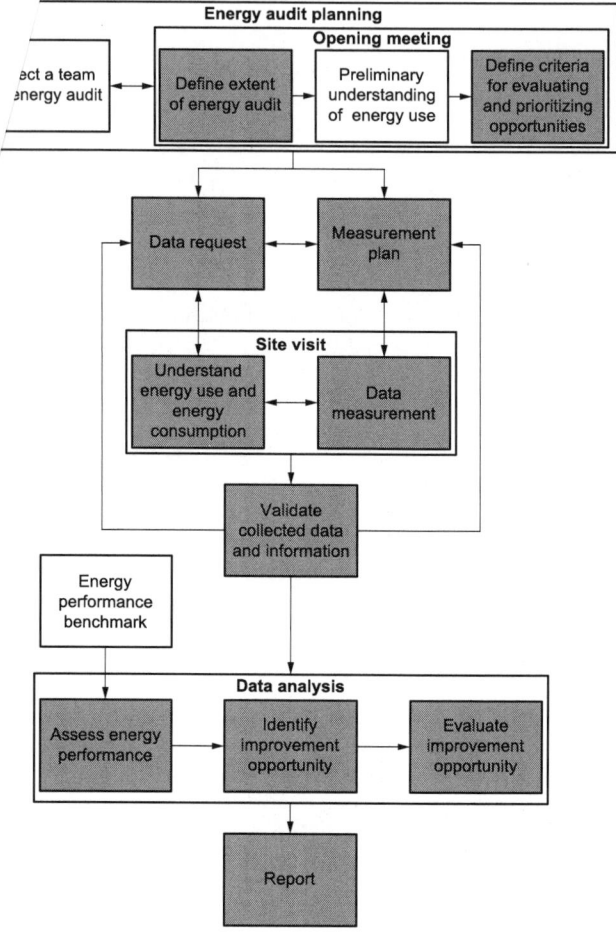

Figure 5.1: Energy audit process

Planning the energy audit

As described in Chapter 4, among the very first tasks during an energy audit is to select the appropriate scope, boundaries and an energy auditor. There may need to be some iteration within the energy audit planning steps before the scope and a competent team (internal, external or a combination of both) can be finalized. A checklist of the items to be discussed and agreed between the organization and the energy auditor is as follows:

- Why does the organization want to carry out an energy audit? Some examples are:

Planning the energy audit

    a) to identify the energy users and to quantify their energy consumption within the organization;
    b) to use an energy audit as the first energy review document required by an ISO 50001-based energy management system; and
    c) to reduce energy consumption while maintaining the same level of business activity.

Costa Coffee's coffee-roasting facility in Lambeth, London, trebled its production between 2009 and 2013. The existing capacity of gas and electricity was limiting how far it could expand the production throughput at Lambeth. As such there was a real need to increase production without an increase in energy consumption. Costa Coffee replaced the existing roasters with energy-efficient models, replaced the air compressors, relamped the existing lighting with LED lighting, reviewed its roaster processes to identify other energy-saving opportunities, and implemented an employee engagement process. It achieved a 32 per cent energy reduction per ton of coffee compared to 2009 and avoided the need to upgrade the existing energy supply capacity or to shift production out from its heritage site.[90]

- What does the organization hope to achieve (or gain) from the energy audit? Some examples are:
    a) to evaluate the feasibility and viability of specific opportunities for improvement, e.g. steam traps, high-efficiency motors, the application of VSDs and heat pumps;
    b) to identify and quantify a portfolio of opportunities to save energy by 'X'%; and
    c) to identify and quantify opportunities to reduce energy consumption in energy systems 'Y' and 'Z'.

PricewaterhouseCoopers' (PwC's) head office in Charing Cross, London, is a nine-storey building with 42,000 m^2 of office space. Built in the early 1990s, the building and facilities were at their end-of-life stage. PwC set a target to refurbish the workspace and facilities located in the basement, roof and terraces to a BREEAM outstanding rating while the building remained in occupation. The refurbishment achieved a BREEAM rating of 96.31 per cent – the highest rating to date for new builds and existing structures.[91]

- What energy use is to be audited? Should the energy audit cover the whole site, all energy-using systems or a specific energy-using piece of equipment? If a whole building or a whole process is not included, who shall evaluate (and how) the interaction between the energy users in the scope and those outside the scope of the energy audit?
- What level of detail is expected from the energy audit? For example:
    a) the energy reduction calculations should be based on measurements and have an accuracy of ± 'A' per cent upon implementation, and calculations should be appended to the end of the report for record-keeping purposes;

Chapter 5   The processes of an energy audit

   b) the total project cost should be based on equipment costs only or on the total implementation cost;
   c) the total project cost should be accurate to ± 'B' per cent for capital sanction; and
   d) all opportunities shall be based on existing measurements and be supplemented by additional measurements, as necessary.
- How long does the energy auditor have to complete the energy audit? For example: (1) the energy audit should be completed and a draft report be available for comment by 'C'; or (2) the energy audit should be completed within 'D' weeks.

Newham University Hospital – Barts Health NHS Trust's energy efficiency programme ('RE:FIT') had an eight-week site investigation from the commencement date. The opportunities for improvement were to achieve a minimum of 6 per cent savings from year one with a maximum payback period of seven years, be implementable within 16 weeks from the contract award, all installations needed to conform to the relevant Health Technical Memoranda (HTMs) and Health Building Notes (HBNs), and it was not to exceed a capital cost of £450,000 (inclusive of VAT).[92]

- How shall the energy auditor evaluate opportunities for improvements identified during an energy audit?
   o What cost basis shall the energy auditor use to calculate the financial benefit arising from the improvements? For example, (1) use an 'E' rate for electricity and a 'F' rate for natural gas; or (2) use a 'G' tariff (or rate structure) for energy costs.
   o If there are other benefits arising from the opportunity, e.g. water, effluent, waste and maintenance reduction, or an improvement in quality and yield, how shall these benefits be captured and quantified?
   o What method should be used to evaluate the financial viability of the identified opportunities? Some examples are simple payback, discounted cash flow with 'H' discount factor, and life cycle costing with 'I' energy cost escalations every 'J' years.
   o How should the opportunities be prioritized for implementation? Some examples are to:
      a) rank opportunities based on the shortest simple payback to the longest simple payback;
      b) rank the opportunities based on the CO_2 avoided;
      c) group the opportunities into 'K' categories or investment profiles.
   o Any other criteria.

adidas Group has a $2 million greenENERGY Fund that assesses and funds energy-saving projects. The fund assesses energy reduction projects from across all of adidas Group as a portfolio of projects, to generate a 20 per cent annual return on capital. If a project falls below the 20 per cent return threshold, it competes with other projects in the group on a ton

CO_2 per dollar-invested basis. This way, projects with a high return help to finance projects with lower returns but which have high CO_2 savings. All funded projects are also published internally and become sources of inspiration for other facilities.[93]

- What resources and support will the organization provide the energy auditor with in order to carry out and complete an energy audit? How should the organization and the energy auditor communicate with each other? If there are unforeseen circumstances (or opportunities) that arise during the energy audit, how can the energy auditor and the organization discuss and agree on a way forward?
- What data are relevant to the scope of the energy audit? Are they available? Some examples might be: production volume, building occupancy trends, drawings, historical energy consumption, manuals and other technical documentation.
- Other relevant information such as, regulatory or other constraints, recently completed projects, planned projects, outsourcing facilities management and changes or other considerations that may have an impact on the energy audit.

It is important that all of the above are discussed and agreed between the organization and the energy auditor. These items should, as far as possible, contain details proportionate to the breadth of information available, and the type of energy audit to be undertaken. Sometimes, it may take several meetings to define a clear scope of works and energy audit criteria. Once these have been agreed, it allows:

- the remainder of the energy audit activities to be planned;
- a competent team of energy auditors to be chosen based on the scope of the energy audit.

Choosing energy auditors with the right knowledge, skills and experience adds value to the organization, as they identify energy reduction opportunities and build a business case for their implementation, which facilitates the organization to make good investment decisions. Chapter 7 shows a simple framework by which a competent energy auditor, meeting the energy audit scope, can be defined and determined.

Roles and responsibilities

As mentioned in Chapter 2, energy audit standards are a good source of information on the various generic processes of an energy audit. When conducting an energy audit, all of the processes are carried out by a person or a group of people. Human interaction is an integral part of an energy audit. While energy audit standards such as ISO 50002 describe the processes as a task for the energy auditing team, the organization,

Chapter 5   The processes of an energy audit

too, has roles and responsibilities in order to facilitate accurate and meaningful findings. Some of these roles and responsibilities are:

- determining the need for an energy audit;
- defining the objectives of an energy audit;
- defining the scope of an energy audit and the criteria for assessment;
- selecting a competent energy audit team that is proportionate to the scope of the energy audit and the level of detail required;
- discussing and coming to an agreement on what tasks and resources are to be provided by the organization and by the energy auditor;
- approving and implementing the energy audit plan and schedules;
- providing the energy auditor with the agreed resources. This may include time, human resources, access to the relevant energy systems, and access to the relevant information and data;
- informing the relevant employees within the organization about the energy audit and directing them to assist in the energy audit activities.

Opening meeting

At the end of the energy audit planning, some time may elapse before the actual energy audit begins. There can be many reasons for this time lapse; they can range from taking time to gain the necessary approval from higher management in the organization, and to collect some agreed preliminary data or information for the energy audit, to juggling organizational priorities to organize resources. The next logical process is to have a so-called 'kick-off' or 'start-up' meeting. This is normally a face-to-face meeting but can sometimes be a telephone call, or a telephone-, video- or web-based conference between the organization and the energy auditor.

The purpose of this meeting is to, among others, remind everyone involved, and to brief those new to the energy audit, about the agreed works, roles and responsibilities of the different parties, obtain an update since the last discussion and agreement, and confirm visit schedules. It may also be an opportunity to collect site-specific information and plans, e.g. unusual operating conditions, or planned or unplanned maintenance, highlight or seek clarification on the provided data or request additional data.

It is also prudent to discuss the need for regular contact and progress updates between the organization and the energy audit team. This could be a regular review of the works, discussions about the data or measurements provided, or discussions on possible opportunities for improvements. These touchpoints can help to develop the concepts for improvements and flush out non-practical solutions or 'tried and failed' opportunities.

For energy audits agreed at a corporate level, the kick-off meeting may also be the first time the energy audit team meets the site-level employees who will be acting as the local liaison, arranging health and safety requirements, and access to the site and energy use to be audited.

Data and information

An analysis of 225 commercial properties in London with a total floor space of 2 million square metres with an Energy Performance Certificate rating of between A and G showed that there is very little correlation between a building's design energy performance (as measured by its Energy Performance Certificate) and actual energy consumption (as measured by its DEC).[94]

As such, using real energy and other data to quantify energy consumption and energy efficiency is important to arrive at a realistic and achievable energy reduction. During an energy audit, the energy audit team analyses the energy performance of the specified scope and comes up with recommendations for improvement. It involves collecting real operating data from the organization and, if necessary, taking additional on-site measurements, and then analysing and making sense of these data.

Data to be collected

The following shortlists a range of information an energy auditor is likely to request:

- *a list of known energy users within the scope of works.* If the scope is small, this could be a list of all the equipment within the scope. If the scope is a system, e.g. a cooling tower, through to its distributions and the end users of the utility, this could be the cooling tower and all of its ancillaries, and a list of all users that consume cooling water. If the scope is the full boundaries of the facility or building, a complete list of energy users will shed light on the complexities of the energy audit and help the energy auditor to frame the scope and its interdependencies, and potential knock-on effects. It may help to break down large boundaries and scopes into smaller scopes, where the energy audit could be prioritized or contracted to a different team;
- *a description of the process or energy flows to and from the energy users.* This could be a description, a sketch or a walk around with the energy audit team to show how the energy use operates, and how the energy is generated and distributed to each energy user. It could also be a collection of operating manuals, commissioning documents

Chapter 5 The processes of an energy audit

and maintenance records. If the organization utilizes the success map described in Chapter 3, this can also be used;
- *energy consumption data and their corresponding variables.* These could be data from utility invoices, half-hourly data from the utility company or data from sub-metering. More importantly, the matching variables are necessary. These could range from the temperature set points for the air conditioning system and the occupancy rates to the detailed production rates, the weather data from the closest official weather station such as the UK Met office or airports, the quantity of electronic data processed and the production mix information. As the saying goes, 'garbage in, garbage out'; very frequently, energy consumption data and/or variables do not provide consistent and repeatable information because they do not tally. A good set of energy consumption data and their corresponding variables are important to quantify the feasibility of the various energy reduction opportunities;
- *instruments that give energy information, its location and how the organization uses the measured information.* Very frequently, information relating to the energy use within the scope of work is available. It just may not be in the form of an energy meter. The information may be in the form of flow, temperature, pressure and time. These data may be recorded in local gauges, programmable logic controllers (PLCs), SCADA systems, control panels, HMIs, BMSs and DCSs. The energy audit team normally prefers that these are available electronically. The energy audit team uses these data alongside other information to analyse the historical energy consumption and energy performance and, if required, to forecast future energy consumption and energy performance. If the data and information are in electronic format, it would speed up the analysis. If these are not available, a trend chart or tabular data is also suitable;
- *current, or a reference, energy cost and a preferred method to evaluate the financial viability of energy improvement opportunities.* This information is important for using as a cost basis and as a financial method to evaluate opportunities for improvements. For example, an organization may prefer the use of discounted cash flow (DCF) with a fixed costs and discount factor. Another organization may use simple payback (SPB) at a first pass and use discounted cash flow with a weighted average cost of capital (WACC) for capital sanction. Organizations will have their own preferences regarding financial analysis and these are dependent on their business environment, risk, cash reserves and their debt and profit structures. Other common tools for evaluation are internal rate of return (IRR), net present value (NPV), life cycle costing (LCC) and real options;
- *plans of the organization that may affect energy use, energy consumption and energy performance.* These could range from planned expansions or contractions, changes in production volume and the introduction of new products, to major retrofits and asset

Data and information

replacement of equipment or systems. It is capitally inefficient to implement an improvement opportunity with a return on investment of five years, if it is to be removed or replaced the following year. That investment would have to be written off. Therefore, it is important that the energy audit team considers this information when assessing the technical feasibility and economic viability of energy reduction opportunities.

If real operating data is not available, the energy audit team could use benchmark figures or estimates, or assume certain operating practices, to arrive at an estimate of energy consumption and/or energy savings. This would certainly be beneficial for type 1 energy audits, to identify opportunities in a broad-brush and prompt manner. For an energy audit outcome to be accurate, more operating data is essential. The use of real data should form the basis for type 2 and type 3 energy audits.

Data to be measured

With all good intentions and preparation prior to an energy audit, it is highly unlikely for an organization to have all of the requested information to hand. The energy audit team is likely to need additional, on-site data measurements.

Before proceeding with on-site work, it is best to spend some time planning the type and resolution of any additional data needed, and to assess whether the intended measurement is practicably obtainable. It could be that the measurement point is located at a high level or in a tight or confined space, or that the configuration of the pipe installation makes readily available measuring devices useless.

In some installations, e.g. large building blocks or large distribution networks, it may not be practicable or cost-effective to measure every data point simultaneously. In this case, measuring and gathering a representative sample, or prioritized sampling, is necessary.

If additional data is necessary, the energy audit team works alongside, and in conjunction with, the organization, to identify a range of additional data and measurements, and how the data can be interpreted in subsequent analysis. A good practice in planning data measurement is as follows:

- list the measurement points, the location of additional measurements and the feasibility to take measurements at the identified locations;
- assess the appropriate measurement technique: e.g. use of bucket and timer, portable measurement devices, use of intrusive measurement device at existing locations such as binder points, and/or observing the BMSs or the control room monitor that is not recorded and trended;

Chapter 5   The processes of an energy audit

- identify the conditions when the appropriate measurements are to be taken such that the readings are representative of the organization's activities;
- determine the frequency for each measurement: e.g. every second, every minute, every 15 minutes or every 30 minutes;
- assess the required accuracy and repeatability of the measurements. If the measuring device has a calibration requirement, check that the measuring device is in operating condition and that the calibration is in date;
- discuss and decide whether the organization's personnel, the energy audit team or subcontractors will carry out the measurements. If there is a need for the organization to simulate different operating conditions, this should also be discussed and agreed beforehand;
- if additional data measurement is not feasible, the energy audit team discusses it with the organization and uses other methods to arrive at the data required. This may be via educated assumptions, estimations, comparison from a like-for-like time frame, benchmark from similar applications, etc.

Data analysis before site visit

As suggested by Chapter 3 and Chapter 4, the energy audit team can now begin to compile and collate the data provided by the organization. Carrying out basic analysis before the site visit helps to identify some opportunities early on and also helps to frame areas of investigation. Some common pre-site visit analyses involve:

- plotting the half-hourly energy data to review the patterns of energy consumption, and spotting unexpected energy consumption and reoccurring patterns and trends;
- plotting the energy consumption over time to observe and spot reoccurring patterns and trends. It may be that the energy consumption rises during periods when no one is at the organization or there may be some automated control systems calling for energy consumption. As another example, it might be that the security personnel does their security walk around and forgets to turn off the energy-using equipment. It could also be that the organization expects only one machine to be running but the energy pattern suggests more than one is running;
- plotting the energy consumption versus several known variables that affect energy consumption. Some of them might be ambient temperature, humidity, production volume, occupancy, daylight and mileage. Using these analyses, the energy auditor can begin to assess whether the energy consumption varies with the parameters that are relevant to the operation of the organization, i.e. business activity. If there is a correlation, is the correlation a loose fit or a tight fit?

These analyses also allow some user set points or equipment operational data to be deciphered and identified prior to visiting the site;
- the energy audit team reviewing the site documentation and familiarizing itself with the site, and researching into specific areas of improvement that may be relevant to its energy use, energy consumption and energy efficiency.

These pre-site visit analyses also serve as conversation starters with the personnel from the organization. They facilitate conversations that probe and generate an understanding of the organization's energy use, and that begin to investigate the reasons for any deviations or departures from expected energy consumption and/or energy efficiency.

Observing the organizational activities and physical operating conditions

The next logical process in an energy audit is for the energy audit team to observe how the machine and the operators work within the scope of works. This is where the majority of energy data, observations and measurements are made. This includes:

- confirming the installation is as described during the data collection phase;
- observing and taking additional measurements of the energy use during start-up and shutdown, normal operation and idle periods, and operator interactions;
- confirming actual operating parameters and observations, e.g. from equipment nameplate data, local temperature and pressure gauges, HMI systems, BMSs or the site's metering systems.

The energy audit team may also need to change the measurement requirements to ensure the measured data are representative of the actual operating conditions, or to make additional measurements that were not foreseen during the planning stages or due to new observations at the site.

When the energy audit team is on-site, one useful source of information is interviews with key stakeholders, in particular, those directly engaged with the operation of the energy use, processes and/or machines. This helps the energy audit team to verify the information provided, and to obtain other essential information for later analysis. Other people that can provide good insight into the energy use are the:

- site/plant/building manager;
- energy manager;
- maintenance/engineering manager;

Chapter 5 The processes of an energy audit

- projects manager;
- finance manager;
- operations/technical services manager;
- production manager.

If the energy audit team was provided with information and data requested during the planning step, it may have analysed the data and identified preliminary ideas for improvement. The site visit also serves as an opportunity to check their applicability.

Analysing opportunities for improvement

After completion of the site work and the data collection exercise, the energy audit team carries out a high-level check and balance of the information and measurements that have been collected. This is to ensure that the measurements are representative of the conditions observed during the site visits, and that they tally with the information about the operational pattern of the energy use.

If the data and/or measurements do not tally, or there is an unexplained anomaly where additional information or measurement is necessary, the energy audit team will arrange to revisit the site to gather additional data and information. If the data and information gathered is sufficiently representative to meet the requirements and objectives of the energy audit, the energy audit team begins to analyse the data and information.

Depending on the scope of works, this could be an audit of the energy performance of a single piece of equipment, of the system or of the whole facility. The evaluation of energy performance forms the basis for identifying and evaluating the opportunities for improvement. Because of this, it is important for the energy auditor to use established, transparent and traceable calculation methods, and document and verify that assumptions made are applicable and valid for the observed data and information.

Assessment of the energy performance may also include trending of historical energy consumption, a breakdown of the energy consumption by energy use to show the proportion of energy consumption of various pieces of equipment within the scope, and analysis of the energy consumption against the known variables. For licensed processes and energy audits with a complex scope, the energy audit team may also utilize computer simulations and preprogrammed software to facilitate the analysis of energy data and information.

A close examination of the actual energy performance data and comparing it with either the minimum energy consumption to meet the company's demand (as described in Chapter 1) and/or the benchmark energy performance gives a gap for improvement.

Analysing opportunities for improvement

Some commonly used benchmarks for buildings are the Department for Communities and Local Government's (DCLG's) Conservation of Fuel and Power Approved Documents: L1A,[95] L1B,[96] L2A [97] and L2B.[98] These documents give information on the standards of insulation, windows, roof and other parameters that are required to achieve the latest regulatory requirements. A more stringent requirement, i.e. a building that uses significantly less energy, would be the Passivhaus [99] (new building) and EnerPHit [100] (building refurbishment) standards.

Organizations operating in a regulated environment such as hospitals have their own code of practice. For example, *Heating and Ventilation Systems – Health Technical Memorandum 03-01: Specialised Ventilation for Healthcare Premises* [101] provides guidance on the air ventilation rates and temperatures for different areas in a hospital. For other buildings, the Chartered Institution of Building Services Engineers (CIBSE) *Guide A* and the American National Standards Institute (ANSI)/American Society of Heating, Refrigerating and Air-Conditioning Engineers (ASHRAE) *Standard 62.1* provide good sources of information on the minimum acceptable ventilation.

For commercial and industrial equipment, the benchmark is normally the equivalent of new equipment of the various equipment manufacturers. A good single source for energy efficiency benchmarks in commercial and industrial equipment is the Energy Technology Criteria List [102] from the UK government's Energy-Saving Enhanced Capital Allowance (ECA) scheme. This list is updated annually and provides the minimum energy efficiency values that are acceptable, along with the calculation methods to be used.

For a scope of work involving a system, for example, a hot water system, the minimum energy consumption the whole system needs in order to meet the energy demands of the end users is the benchmark.

The next part of the analysis is to identify opportunities for improvement based on the data and the actual configuration, or layout, of the energy users within the scope of the energy audit. There is some degree of freedom regarding views that the energy audit team will take into account. Typically, these are based on the energy audit team's experiences, the condition of the equipment, the age of the equipment and good engineering practices. As such, two separate energy audit teams working on the same energy users often arrive at a different technical solution (see Chapter 7).

In general, opportunities for improvement involve removing the need for energy use, minimizing energy consumption and ensuring high efficiency. They can be classified into one, or a combination of, the following:

Energy Audits

Chapter 5 The processes of an energy audit

- *improving energy management practices.* Examples include introducing metering and/or sub-metering, putting in place a better energy monitoring technique, and implementing an energy management system;
- *improving operational practices.* Examples include incorporating energy-efficient behaviour and energy efficiency in standard operating procedures and ways of working, minimizing energy waste, gaining better control of process variability, improving the equipment layout and/or set points to minimize idle times, controller tuning, production scheduling, minimizing waste, operating the most efficient machine to suit the demand, and applying and/or replacing poor insulation;
- *improving maintenance practices.* Examples include minimizing unplanned downtime, incorporating root cause analysis in equipment failures, ensuring maintenance is carried out according to equipment suppliers' manuals, minimizing leaks, implementing routine condition monitoring, and heat exchanger cleaning;
- *utilizing energy-efficient behaviour change programmes.* Examples include energy-efficient training, energy awareness campaigns, specific chain of communication events or campaigns, non-monetary motivational techniques and financial-based incentives;
- *modifications and retrofits of energy-efficient products.* Examples include the application of condensing boilers, VSDs and energy-efficient lighting.

In 2010, the University of Hertfordshire, as part of its energy efficiency efforts, set a goal to refurbish one of its data centres located in Hatfield. Four key opportunities identified were to:

1) consolidate the number of servers from 23 to 15 by the use of VMware virtual servers and Sun containers, to increase the utilization of the servers;
2) widen the operating window for acceptable air quality (for temperature and humidity) inside the data centres, according to ASHRAE's Thermal Guidelines for Data Processing Environments 3rd ed.[103]
3) utilize the hot reject heat from the data centre to provide hot water in the adjacent offices; and
4) utilize free air cooling to cool the data centre for 86 per cent of the year.

These resultant opportunities allowed the university to reduce its data centre power usage effectiveness (PUE) – an energy performance indicator for data centres – from 2.2 to 1.2, giving energy savings of £40,000 per year.[104]

The energy audit team then develops these opportunities into concepts and preliminary designs. The purpose of these concept designs is to allow implementation costs to be developed, and the feasibility of the designs

Analysing opportunities for improvement

and the economic viability of the identified opportunities to be assessed, based on the agreed economic criteria established during the planning stages of the energy audit. Depending on the scope of works and the level of detail required from the energy audit, the accuracy of implementation costs, and hence their return on investment, varies.

For a type 1 energy audit, the energy savings and investment figures are rough and ready. For zero-cost and low-cost opportunities, it is suffice to progress these opportunities. For type 2 energy audits, the investment required is usually based on the energy audit team's experience or on the major cost items plus a percentage for the installation works. The typical percentages are 100 per cent for any civil works required, 100 per cent for mechanical works, 100 per cent for electrical and instrumentation works, and 5 per cent to 25 per cent for project management. Some energy audit teams use cost index tables for arriving at investment costs. While this is acceptable, these figures should be taken with a 'pinch of salt' because:

- with the speed of technological advancement, the capital cost of major equipment is constantly revised, with the cost of equivalent specification machines or equipment coming down a lot faster than the tables are updated;
- the cost of major equipment depends on the major equipment manufacturer. For example, when purchasing a high-efficiency motor or a VSD, purchasing it from a reputable brand would be more expensive than purchasing it from a generic brand. However, the former come with other features and/or benefits that, perhaps, may not be available in other brands;
- the cost of major equipment also depends on from whom the equipment is purchased. If the machine or piece of equipment is to be purchased from a supplier with many handling agents, it could have a higher cost than if it were purchased directly via the main distributor;
- lastly, the cost of the major equipment and the cost of the project also depend on how the equipment is to be installed. Using the implementation of a VSD as an example, some organizations allow the use of a built-in controller on the VSD to control the speed of the motor – all the set points are preprogrammed into the VSD directly. Some companies prefer the speed control to be handled via an externally located BMS or a centralized control system. In the latter case, there would be an increase in electrical and instrumentation costs to allow the energy savings to be realized.

For a type 3 energy audit, the costing information may be based on draft concept drawings, general arrangement drawings, constructability and operability assessments, health and safety considerations, environmental considerations, and detailed scope of works for the various trades. It is also prudent to have detailed and quoted costings at this stage as they

Chapter 5   The processes of an energy audit

form the basis for project costings and capital assessments. The costings for smaller items such as professional fees, measurement and verification and contingencies may still be based on a percentage of the quoted prices.

At this point, some energy audit teams review the findings and opportunities before categorizing and ranking them. Discussions at this stage allow both the energy audit team and the organization to examine the feasibility of the opportunities and to identify any non-energy improvements that may arise from those same opportunities, e.g. maintenance reduction, removal of equipment bottlenecks, improvement in working conditions, improvement in quality and reduction in waste.

The discussions may also uncover potential (positive and/or negative) interactions and knock-on effects from the opportunities. It may transpire that a reduction in energy consumption within the scope of the energy audit leads to an increase in energy consumption outside the scope of works. For example, a reduction in steam flow in a distillation column could lead to a reduction in the purity of a mixture, which has a knock-on effect of increasing the energy consumption in subsequent energy users, increasing the amount of quality rejects or generating off-specification trade effluent, where the organization could be fined.

As another example, a 100 W incandescent light bulb emits 5 W as light and 95 W as heat. A 10 W LED light bulb emits 5 W as light and 5 W as heat. When calculating energy savings from replacing the incandescent light bulbs with LEDs, using LEDs gives electricity savings of 90 W. However, there is another component frequently missed in the calculation: a 95 W reduction in heat. During summer, the 95 W of heat would otherwise need to be removed by the existing cooling system, which would reduce the energy savings. During winter, an additional 95 W of heat would need to be added back into the office by the heating system to make up for the reduced heat inside the building. The heat energy to and from offices also needs to be considered when doing a lighting replacement project.

Discussions between the energy audit team and the organization also provide an opportunity to validate any assumptions made during the assessment, and to correct and refine the assumptions and assessments. One of the most frequently used assumptions that has significant consequences if it is proven wrong is the use of averages. This concept is collectively known as 'the flaw of averages' [105] and manifests itself in several guises:

- *using average values in energy costs.* For many organizations, the cost of electricity consists of a capacity (demand) charge, a higher, day consumption (peak) charge and a lower, night consumption (off-peak) charge. However, the cost of electricity is often calculated using a cost figure that is the capacity charge normalized over 30

days plus an average of the daytime and night-time electricity rates. If the energy reduction occurs at night, using the average cost of electricity will overestimate energy savings. First, this is because, in both cases, the capacity charge does not change – it is the consumption that changes. Secondly, within the consumption savings component, it is the off-peak component that is reduced, not the peak component. If the energy reduction occurs during the day, using the average cost of electricity will underestimate savings;
- *using average values as a baseload.* Taking the weather as an example, the temperature varies throughout a 24-hour period, and also from month to month. Figure 3.8 shows this over a full year. Using an average temperature as a baseload and selecting the more energy-efficient heating replacement based on the average load will result in the heating replacement not able to meet the heating demand during winter when temperatures drop below the average temperature;
- *using average values in energy consumption.* Some energy-using machines and equipment have multiple modes of operation. In certain modes of operation, one of these machines may consume very little energy, while in other modes of operation, it may consume considerably more. Depending on the number of hours the machine spends in each of the different modes of operation, using average energy consumption for the machine will result in either a significant overestimation or a significant underestimation in energy consumption. This will lead to a significant overestimation or underestimation of the energy reduction opportunity and effort in calculating the return on investment.

Another category of assumptions that are more often wrong than not is the use of load factors, i.e. the capacity utilization of electricity consuming equipment and machines, in particular, motors, pumps or fans. Many assume that a motor, pump or fan will operate at capacity between 70 per cent and 80 per cent of the time. A survey by Robert Hoshide found that less than 25 per cent of motors, pumps and fans installed operate at more than 60 per cent of their rated capacity.[106] In this case, making an assumption of 70 per cent to 80 per cent utilization factor will overestimate the energy savings for more than 75 per cent of the time!

The disclosure and/or description of inputs into the energy analysis is useful and allows the organization to check if these inputs are representative of the operation's conditions. Of particular interest might be information about how the energy auditor uses information to analyse non-routine operations, malfunctioning (but have not failed) equipment, relationships with interdependent machines and equipment, physical characteristics such as building height, shading and other microclimate effects, and basis for cost estimates.

Finally, the identified opportunities are categorized, ranked and prioritized according to the method agreed during the planning stages of the energy audit. An indication of the project schedule and the high-level method for verifying the energy reductions may also be drafted.

Recalling the RBS Group example introduced in Chapter 4, the subsequent steps by the RBS Energy Team were:[107]

- *design.* The analysis of energy data and control systems revealed many different and, at times, conflicting control philosophies, leading to boilers and chillers operating simultaneously and unnecessarily, thus consuming energy without providing any real benefit. The building occupancy patterns were established for each zone and floor to redesign the control and operating philosophies. A number of defective plant items were also established and prioritized for maintenance. Free air cooling was also assessed and its feasibility established;
- *implement.* All of the opportunities above were implemented and incorporated into the MOPs' BMSs;
- *sustain.* Following the recommissioning of the ventilation and air conditioning systems, energy reductions were verified. This has saved the RBS Group 12.4 GWh of electricity and 4.7 GWh of natural gas, giving an overall cost saving of £1.2 million per year.

Energy audit reporting

This step of the energy audit pulls together all of the information from the energy audit: from planning meetings through data collection phases to the site work and data analysis work. In essence, this step provides an opportunity for the energy audit team to (1) review and check that energy audit output meets the scope of works, aims and objectives, and (2) provide the organization with an insightful report and recommendations.

An energy audit report is written with an audience firmly in mind:

1) use the executive summary for busy executives interested in the bottom line;
2) use the main body of the report, perhaps with the aid of carefully selected graphs and charts, to inform a middle and/or junior management; and
3) leave the bulkier information to the appendices.

The use of technical jargon and words with multiple meanings, such as high-efficiency, low carbon, premium, green, world-class and value-added, should be avoided.

Energy audit reporting

The exact content and details included in the report are dependent on the scope and type of energy audit. As a guide, the report should cover the following topics:

- an executive or management summary;
- a description of the energy audit, the scope of works, the collection of information and measurement exercises;
- an analysis of the energy performance and energy performance indicator(s) of the scope, based on measured and collected data;
- the basis for the energy cost-saving calculations, estimates and assumptions used in the report, and the criteria for ranking opportunities for improving energy performance;
- the opportunities for improving energy performance, including:
  a) any assumptions used to quantify the energy reductions and cost of implementation;
  b) the accuracy and limits of accuracy of the studies;
  c) quantify, with assistance from the organization, any non-energy benefits from the identified opportunities and potential interactions with other proposed opportunities, or with other energy users;
  d) additional actions necessary to implement the opportunities;
  e) a cost–benefit analysis using the agreed methods;
  f) the ranking of opportunities based on agreed methods;
  g) recommendations for improvements; and
  h) outline a method to verify the savings once the recommendations have been implemented.
- an energy audit conclusion.

Many energy audit teams also include a background section in the energy audit report. This section usually details general information about the organization and about the energy auditor, and provides a detailed description of the scope, a statement about confidentiality and other relevant information, such as legal requirements and other requirements applicable to the energy audit or to the organization.

A good energy audit team is also open and transparent in terms of describing the proposals in detail rather than giving minimal information in a bid to 'protect their intellectual property'. This becomes very useful as many reports are actually not read by top managers themselves.[108] For many, top managers get to know about the results of an energy audit through managers doing a similar presentation and making a recommendation to the board.

Closing meeting

Lastly, the energy audit team is in a position to present the findings to the organization (its observations, analysis and recommendations), address any clarifications required and questions asked, and close the commercial arrangement.

Some energy audit teams have a meeting with the organization with a draft report, carry out similar activities to the above, and then incorporate the discussions and additional information before finalizing the report and concluding the commercial engagement.

Chapter 6   Using energy audit processes to maximize energy savings

When opportunities for improvement are implemented in a specific sequence, organizations can maximize the energy savings and minimize the capital investments. Chapter 1 introduced this sequence as the energy maturity model. The model proposes that the easiest and most obvious method is to prioritize opportunities that are the cheapest to implement and, therefore, are the most cost-effective measures, i.e. the 'low-hanging fruit'.

For many energy users, good housekeeping opportunities are: good maintenance, turning things off, improving insulation, and reducing waste, leaks, idle time, production rate losses and the incidences of untended taps and hoses. This would require the organization to engage with its employees.[10] Success in instilling this zero tolerance policy towards energy performance minimizes the non-value-added use of energy.

The next category – control systems – tightens the controls of existing processes and utilities, requiring them to operate closer to the control limits. Some examples are:

- introducing better operating conditions for existing energy users;
- introducing temperature dead bands in air conditioning systems;
- matching the most efficient machine with actual demand;
- using preventive maintenance and condition monitoring to predict and prevent equipment failure;
- reducing excess usage;
- increasing cycles of concentration;
- better control of water treatment systems;
- using high-efficiency motors and VSDs;
- using information from control systems to monitor equipment performance.

The next level of energy maturity is achieved by implementing simple modifications and/or refurbishments. These may include the reuse and

---

[10] For an introduction into employee engagement and behavioural change, please see Appendix A.

Chapter 6 Using energy audit processes to maximize energy savings

recycling of thermal energy, ideally, local, direct use without the need for tanks, pipework and pumps. Some of the further potential modifications and/or refurbishments include installing waste heat recovery boilers, preheaters and economizers, redesigning to use less energy, e.g. creating a close loop condensate system, and using an energy-efficient component in a machine. One of the drawbacks in heat recovery is found in both the source of heat and its user: the heat source has to generate heat and the sink has to be using energy at the same time. If either one is out of sync, the potential for energy savings is reduced.

Further integration of energy use is the next natural progression for higher energy savings, and includes:

- recovering energy from one energy user to be reused by another energy user;
- thermal pinch analysis;
- process intensification, optimization and enhancement; and
- plant de-bottlenecking and uprating.

The highest available form of energy maturity comes from a step change perspective by commissioning an alternative and more modern process. This step change could include installing and operating a CHP plant, refitting a process with the latest designs and improvements, or applying dynamic simulation and predictive controls.

Table 6.1 to Table 6.12 show some examples of energy maturity stages for several common energy users. An introduction to the people aspects in energy management and behaviour change can be found in Appendix A.

Chapter 6 Using energy audit processes to maximize energy savings

Table 6.1: A shortlist of improvement opportunities for a steam system

Housekeeping	Control	Modification	Integration	Alternative
Isolate idling boilers.	Insulate exposed steam and condensate pipes.	Use an air preheater.	Recover waste heat as a heating source of the standby boiler.	Use hot water boiler instead of steam.
Implement a routine maintenance regime.	Close stack dampers when the burner is off.	Use an economizer.		Burn wastes or biofuel instead of fossil fuel.
Ensure all air trapped in the system is vented during start-up.	Use oxygen trim control.	Use a small boiler for a small summer load.	Use letdown steam turbines to produce lower steam pressures.	Use a CHP plant to augment steam demand.
Monitor water losses from the system.	Stabilize fluctuations in boiler feed water quality.	Recover boiler blowdown heat.	Recover heat from flue gas, engine cooling water, the engine exhaust, low-pressure waste steam.	
Fix steam leaks and condensate leaks.	Minimize boiler blowdown.	Use a deaerator instead of a hotwell.		
Review start-up and shutdown procedures and scheduling to shorten idle time.	Troubleshoot and rectify causes of water quality variation.	Use reverse osmosis (RO) water instead of softened water.	Consider thermo-recompression to minimize high-pressure steam.	
	Return clean condensate to boiler.	Use waste steam for water heating.		
Minimize starting up too early.	Match boiler to load.	Use vacuum pumps instead of steam ejectors.		
Inspect steam traps regularly and repair failed traps promptly.	Operate boiler at lower pressure during standby.	Review the energy use with a view to utilizing		

Housekeeping	Control	Modification	Integration	Alternative
Inspect dirt pockets regularly and ensure they are completely drained.	Stop boilers running when there are no loads.	steam at lower pressures.		
Replace damaged and wet insulation.	Close the crown valve when the boiler is not in use.	Consider automating steam users instead of operating them manually.		
Ensure temperature/pressure/ flow set points on energy users are minimized.	Consider recovery of vent steam (e.g. on large flash tanks).			
Ensure there is minimum fouling and scaling on heat exchangers.	Insulate exposed fittings and pipe supports.			
Set up a routine heat exchanger cleaning regime.	Isolate dead legs and excessive steam traps.			
	Ensure distribution lines, pressure-reducing valves, condensate drainage and steam traps are sized correctly.			
Where applicable, ensure that steam users are operated on automatic mode instead of being operated manually.	Review steam pressure requirements of energy user and minimize pressure on steam generation.			

Chapter 6 Using energy audit processes to maximize energy savings

Housekeeping	Control	Modification	Integration	Alternative
	Find opportunities to minimize steam consumption by energy users.			

Table 6.2: A shortlist of improvement opportunities for a hot water system

Housekeeping	Control	Modification	Integration	Alternative
Fix hot water leaks.	Insulate exposed pipes, fittings and supports.	Use high-efficiency condensing boilers.	Generate hot water on demand, via electricity or steam (if capacity allows).	Burn wastes or biofuel instead of fossil fuel.
Isolate idling boilers.	Isolate dead legs.	Install multiple, smaller boilers rather than one large boiler.	Consider heat recovery opportunities, e.g. from the compressor exhaust.	Use a CHP plant to augment hot water demand.
Replace faulty pressure relief valves and safety valves.	Ensure distribution lines and pressure-reducing valves are sized correctly.	Consider point of use/instant water heaters.	Recover waste heat from steam, condensate or water heat to preheat hot water before it reaches the boiler.	Use heat pumps to augment hot water demand.
Implement a routine maintenance regime.	Stop boilers running when there are no loads (standby losses).	Install economizers.		
In the case of an office building, review and minimize hot water use outside normal working hours.	Stop multiple boilers firing at low loads when one boiler is sufficient.	Recover and return hot water back to the boiler.		
In the case of a building, check that the heat output to the heating element is not blocked or faulty, e.g. blocked grilles, obstructed radiators or clogged air filters.	Stabilize fluctuations in boiler feed water quality.	Install flue dampers to prevent boilers cooling when idle.		
	Minimize boiler blowdown.	Resize the distribution pumps if the existing distribution pumps are oversized.		
Ensure temperature/pressure/flow set points on	Troubleshoot and rectify causes of water quality variation.			

Housekeeping	Control	Modification	Integration	Alternative
energy users are minimized.	Use the differential temperature controller to inhibit boiler starts unless there is a real hot water demand.			
Ensure all trapped air is vented during start-up.	Close and/or minimize bypass throttling valves, attemperation loops, and three-port and four-port control valves, and vary the flow based on demand.			
Monitor water losses from the system.				
Review start-up and shutdown procedures and schedules to shorten idle time.	Review hot water temperature requirements of the energy user and minimize the temperatures at which hot water is generated.			
Ensure there is minimum fouling and scaling on heat exchangers.				
Set up a routine heat exchanger cleaning regime.	Find opportunities to minimize hot water consumption by energy users.			

Table 6.3: A shortlist of improvement opportunities for a cooling water system

Housekeeping	Control	Modification	Integration	Alternative
Ensure temperature/pressure/flow set points on energy users are minimized.	Review cooling water energy use to minimize excessive cooling water temperature set points.	Use two-speed or variable speed drives for cooling tower fan control.		
Verify the water meter readings.	Stabilize fluctuations in cooling water feed water quality.	Stage the cooling tower fans with on-off controls.		
Check for uniform flow of water from cooling tower packing.	Minimize cooling water blowdown.	Install new nozzles and distributors to obtain a more uniform water pattern.		
Check water overflow pipes for proper operating level.	Troubleshoot and rectify causes of water quality variation.	Resize the distribution pumps if the existing distribution pumps are oversized.		
Cover water basins (to minimize algae growth, which contributes to fouling).	Ensure that no other sources of hot air are directed towards the cooling tower drift eliminators.			
Ensure drift eliminators are in good condition and installed correctly.	Control cooling tower fans based on cooling water supply temperature.			
Check and clean plugged cooling tower water distribution nozzles.	If the cooling water distribution pipes are not part of the space			

Chapter 6 Using energy audit processes to maximize energy savings

Housekeeping	Control	Modification	Integration	Alternative
Correct excessive and/or uneven fan blade tip clearance and poor fan balance.	cooling strategy, insulate the pipes to minimize heat pick up.			
Reline leaking cooling tower basins.	Turn off unnecessary cooling tower fans when loads are reduced.			
Shut off and isolate cooling water users that are not in operation.	If the difference in supply and return temperatures is small, review options to increase the temperature difference.			
Establish a cooling tower maintenance regime.	Use the differential temperature controller to inhibit cooling tower fan starts unless there is a real cooling water demand.			
Ensure unrestricted air flows.				
Where two or more cooling towers are sited adjacent to one other, ensure recirculation, or that air starvation does not occur, regardless of wind or building effects.	Close and/or minimize bypass throttling valves, attemperation loops, and three-port and four-port control valves, and vary the flow based on demand.			

Energy Audits

Chapter 6 Using energy audit processes to maximize energy savings

Housekeeping	Control	Modification	Integration	Alternative
Ensure there is minimum fouling and scaling on heat exchangers.	Find opportunities to minimize hot water consumption by energy users.			
Set up a routine heat exchanger cleaning regime.				

Table 6.4: A shortlist of improvement opportunities for a chilled water and chilled glycol system

Housekeeping	Control	Modification	Integration	Alternative
Ensure temperature/pressure/flow set points on energy users are minimized.	Stabilize fluctuations in chilled water and chilled glycol quality.	Review the need for chilled water and/or chilled glycol with a view to changing them to cooling water.	Use cooling water to pre-cool before reaching the chiller.	Use a heat pump to augment the chilling duties.
Check and ensure the glycol concentration is as per design.	Troubleshoot and rectify causes of water quality variation.	Consider the use of water-cooled condensers rather than air-cooled condensers.	If there are considerable quantities of low-grade waste heat, consider using absorption chillers.	
If the chiller is air cooled, ensure there is free access to cooling air for the condenser.	Ensure that no other sources of hot air are directed towards the condenser of the air-cooled chiller.	Temperature differences in supply and return of less than 7 °C should be avoided.		
If the chiller is air cooled, ensure that the hot air exhaust from the condenser is recycled to its inlet.	Find opportunities to minimize hot water consumption by energy users.	Investigate whether the chiller can operate with free air cooling.		
Insulate all chilled water and chilled glycol pipes and fittings.	If the difference in supply and return temperatures of the chilled water or the chilled glycol is small, review options to increase the temperature difference.	Replace old chillers or compressors with new, higher-efficiency models.		
Establish a chiller maintenance regime.		Investigate whether the chiller can operate on variable flow to utilize VSDs.		

Chapter 6 Using energy audit processes to maximize energy savings

Housekeeping	Control	Modification	Integration	Alternative
Do not overcharge refrigerants and oil.	Run the chillers at the lowest operating costs that serve the baseload.	If the difference in supply and exhaust temperatures from the condensers is small, review options to increase the temperature difference.		
Isolate offline chillers and cooling towers.	Review whether refrigerant is the most efficient for chilling duty and application.			
Ensure there is minimum fouling and scaling on heat exchangers.	Review compressor operating profile and choose to operate at the lowest power combination.	If the chillers are not able to operate on variable flow, consider supplying chilled water or chilled glycol to the energy user via a tank.		
Set up a routine heat exchanger cleaning regime.	Review the part-load characteristics of the chillers, and cycling costs, to determine the most efficient mode for operating multiple chillers.	Resize the chillers if the existing chillers are significantly oversized.		
In the case of cold storage, ensure that the doors are closed and the door seals are operable.	Use the differential temperature controller to inhibit chiller starts unless there is a real chilled water or chilled glycol demand.	In the case of open display units, consider the use of glass doors and/or night blinds.		
In the case of cold storage, ensure the goods are arranged evenly.		Use energy-efficient motors for continuous or near-continuous operation.		

Housekeeping	Control	Modification	Integration	Alternative
	Close and/or minimize bypass throttling valves, attemperation loops, and three-port and four-port control valves, and vary the flow based on demand.	Install a control system to co-ordinate multiple chillers. Resize the distribution pumps if the existing distribution pumps are oversized.		

Table 6.5: A shortlist of improvement opportunities for a compressed air system

Housekeeping	Control	Modification	Integration	Alternative
Identify and repair air leaks.	Review energy users to eliminate unnecessary use of compressed air and to minimize compressed air use.	Consider fitting zone isolation valves and interlocks in a large compressed air distribution system.	Consider recovering the heat from air compression for other hot water applications, e.g. to heat a building's make-up air system, to generate hot water or to preheat other processes.	Remove the use of compressed air by energy users.
Turn off idle air compressors.				
Turn off refrigerated and heated air dryers when the air compressors are off.	Review the energy use with a view to comparing and minimizing compressed air pressure use.	Resize the air compressors if the existing air compressors are oversized.		
Ensure that there are no steam and hot water leaks near the intake of the air compressor.	Review the air compressor operating characteristics to determine the most efficient mode of operation.	Consider the use of a steam regenerative, desiccant dryer rather than a compressed air regenerative variant.		
Insulate any exposed steam, hot water and condensate pipes near the intake of the air compressor.	Ensure that the compressed air distribution pipework and receivers are suitably sized.	For energy users, consider the use of electricity driven actuated valves and electronic controls instead of compressed air variants.		
Implement a routine maintenance regime.				
Ensure that manual drain valves are closed after draining and that they form a tight seal.	Take air compressor intake air from the coolest possible location.	Use dedicated high-pressure compressors or booster compressors for		

Housekeeping	Control	Modification	Integration	Alternative
Ensure that the hot exhaust from the air compressor is not recirculated into the air compressor intake.	Review the compressed air quality requirements of energy users with a view to minimizing over-treatment of air quality.	high-pressure air requirements.		
Monitor pressure drops across suction and discharge filters and clean or replace filters as necessary.		Consider high-efficiency motors for compressors.		
		Consider the use of VSDs for variable load air compressors.		
Check for leaking drain valves in compressed air filter/regulator sets.		Consider moving compressed air-operated equipment, such as air knives, air lances, air agitators, blow guns and powder transfer equipment to non-compressed air variants.		
Isolate compressed air to unused energy users or when they are not in operation.				
Install interlocks to areas where compressed air is required intermittently.		Use dedicated blowers for low-pressure applications.		
		Install a control system to co-ordinate multiple air compressors.		
Use other methods for blowing, cooling and cleaning applications.		Change the electric motors to an energy-efficient design.		

Housekeeping	Control	Modification	Integration	Alternative
		Replace standard v-belts with high-efficiency, cogged v-belts.		
		Use automatic, zero-loss drain controls instead of continuous air bleeds.		
		Consider the use of high-efficiency, low-pressure drop filters.		

Table 6.6: A shortlist of improvement opportunities for a ventilation system

Housekeeping	Control	Modification	Integration	Alternative
Isolate areas where ventilation is not required.	Review energy user requirements with the aim of matching supply with recommended ventilation levels, and to minimize excessive air flows and pressures.	For office buildings, consider the use of CO_2 concentration-led ventilation.		Consider the use of natural ventilation.
Turn fans off when they are not needed.		Use smooth, well-rounded air inlet cones for fan air intakes.		
Identify and eliminate ductwork leaks and loose or damaged flexible connections.	Review ventilation requirements when there is a change of use.	Set up ventilation zones based on demand.		
Seal up leaks through doors and windows.	Minimize fan inlet and outlet obstructions.	Consider using interlocks for loading bays.		
Check, clean or replace heating and cooling batteries and filters regularly.	Consider using temperature-controlled strategies to control extract fan speeds in kitchen.	Consider the use of a variable air flow system instead of a constant air flow system.		
Implement a routine maintenance regime.	Use properly sized ductwork with appropriate bends and transitions.	Review ductwork and fittings for low-pressure drop alternatives.		
Check fan blades for erosion; replace as necessary.		Consider installing a building automation system.		
Check that belts are aligned and tightened.	Tune the ventilation control system.	Consider reducing ceiling heights.		

Chapter 6 Using energy audit processes to maximize energy savings

Housekeeping	Control	Modification	Integration	Alternative
Make sure that bearings are lubricated adequately.	Review the BMS to match ventilation demand with occupancy information.	Use professionally designed industrial ventilation hoods for dust and vapour control.		
Check that all dampers and valves are installed correctly and are operational.	Reduce ventilation system operating hours (e.g. at night and/or at weekends).	Upgrade existing filters to high-efficiency, low-pressure drop filters.		
Check that air flows and pressures are balanced and as per design. Identify root causes, if not, and rectify as necessary.	Provide dedicated outside air supply to kitchens, cleaning rooms, combustion equipment, etc.	Review the age and condition of the existing ventilation system with a view to replacing it with an energy-efficient variant.		
In the context of a building, review the possibility to maximize its occupancy.	Consider the use of high-efficiency, low-pressure drop filters.	If a fan is belt-driven, consider the use of low-slip or no-slip belts.		
Ensure that thermostats are located in the right locations.		Consider using a two-speed motor or VSD for variable ventilation demands.		
		Install dedicated ventilation systems for continuous loads (e.g. IT rooms).		

Table 6.7: A shortlist of improvement opportunities for an air conditioning system

Housekeeping	Control	Modification	Integration	Alternative
Prevent unauthorized thermostat adjustments.	Prevent heating or cooling outside working hours.	Use building insulation and solar shading to minimize heat gains and losses.	Consider desiccant drying of outside air to reduce cooling requirements in humid climates.	Use free air cooling.
Check and ensure that heating and cooling batteries are installed correctly and are operational.	Review energy user requirements with a view to defining and/or widening the temperature and humidity control range.	Consider zoning the heating, ventilation and air conditioning systems to minimize energy use.	Where waste heat is available, recover the heat for heating or cooling via an absorption chiller.	Use evaporative cooling in dry climates.
Implement a routine maintenance regime.	In an office environment, introduce a temperature dead band such that the heating does not go above 19 °C and cooling does not go below 24 °C.	Full fresh air free air cooling, based on temperature or enthalpy control, should be considered.		Where climatic weather patterns and energy costs allow, consider the use of heat pumps.
Eliminate simultaneous heating and cooling due to instrument calibration issues, instrument failures and/or valve failures.	Use appropriate thermostat setbacks.	Consider the use of atomization rather than steam for humidification.		
Ensure that the ventilation supply intake is not located in a location where it is potentially taking air from another ventilation system's exhaust.	Minimize low relative humidity demand.	Recover heat from the exhaust air to the supply air before it reaches any heating coils.		
	Carry out the tuning of controls.	Sub-cool the condensate from steam coils.		

Chapter 6 Using energy audit processes to maximize energy savings

Housekeeping	Control	Modification	Integration	Alternative
Insulate and/or replace the insulation on any exposed heating and cooling valves and components.	Relax excessively tight relative humidity set points.	Consider the use of underfloor heating and chilled beams for cooling.		
Reduce humidification or dehumidification during unoccupied periods.	Use morning pre-cooling in summer and preheating in winter (i.e. before electricity peak hours).	In the case of buildings, consider adding a reflective surface to the roofing that minimizes heat gains or heat losses.		
If the building utilizes underfloor heating and chilled beams for cooling, ensure that the floors/ceilings are insulated and are in good condition.	Use spot heating and cooling rather than treating the entire area.	In the case of buildings, consider the use of low-emissivity windows, such as double-glazed windows and triple-glazed windows.		
	Isolate air-conditioned loading dock areas and cool storage areas, using high-speed doors or clear PVC strip curtains.			
In the case of an office building, investigate and resolve issues leading to occupants utilizing portable electric heating appliances.	Reduce unnecessary heat loads, e.g. lighting and IT equipment.	In the case of buildings, consider the use of wall insulation and solar shading to minimize heat gains.		
	In the case of office buildings, designate cooler or warmer zones for occupants in line with the natural heat gain and heat loss profiles of the building.			

Table 6.8: A shortlist of improvement opportunities for a pumping system

Housekeeping	Control	Modification	Integration	Alternative
Repair seals and packing to minimize leaks.	Review energy use to minimize the need for pumping.	Use booster pumps for small loads requiring higher pressures.		
Implement a routine maintenance regime.	Close and/or minimize bypass throttling valves, attemperation loops, and three-port and four-port control valves, and vary the flow based on demand.	Use low-loss commissioning valves.		
Ensure that the pump, its motor and its coupling are aligned.		Consider the use of smooth bends rather than abrupt turns to minimize pressure drops.		
For belt-driven pumps, ensure that the belt is tight.	Review the existing pumping system to maximize opportunities to utilize siphon effects.	Consider minimizing the number of pipe fittings to minimize pressure drops.		
Consider tabulating a list of pumps, their duties and their nameplate data.	Operate pumps at best efficiency point.			
When rewinding a large motor, ensure that the motor is rewound to the highest standard available.	Resize the pumps if the existing pumps are oversized.			
	Where appropriate, replace the motor with the most efficient motor.			
	Monitor and resolve issues relating to voltage			

Housekeeping	Control	Modification	Integration	Alternative
	imbalances, harmonic distortions and poor power factors.			

Chapter 6 Using energy audit processes to maximize energy savings

Table 6.9: A shortlist of improvement opportunities for heat exchange (heat exchanger, vaporizer, evaporator and condenser)

Housekeeping	Control	Modification	Integration	Alternative
Repair all leaks.	Review the energy use with a view to minimizing the flow, pressures and temperatures to be achieved by the heat exchanger.	In a steam-operated heat exchanger, if the duties are permissible, sub-cool the condensate.	For large heat exchangers, consider the use of an absorption chiller or an Organic Rankine Cycle system to sub-cool the condensate before returning it to the boiler.	
Identify and replace faulty pressure relief valves and safety valves.				
For fouling duties, ensure heat exchanger surfaces are cleaned regularly.	Avoid overheating or over-cooling by the heat exchanger.	Consider the use of compact and/or shell and tube heat exchangers.	Consider the use of multiple effect evaporation.	
Replace any damaged and/or wet insulation.	Return steam condensate to the boiler.	In a distillation column, reuse distillation bottoms heat to preheat distillation feeds.		
Implement a routine maintenance regime.	Insulate exposed pipes, fittings and supports.	If the heat exchanger is steam operated, consider the use of thermo-recompression.		
Ensure all instrumentation for the heat exchangers is accurate and repeatable.	Isolate dead legs and excessive steam traps.	Consider automating the heat exchange operation.		
Ensure all instrumentation is operating within normal parameters.	Ensure distribution lines, pressure-reducing valves, condensate drainage and steam traps are sized correctly.			
Ensure heat exchangers				

Energy Audits 147

Housekeeping	Control	Modification	Integration	Alternative
are operated according to procedures.				
Ensure all air is vented from the heat exchanger during start-up.				
If the heat exchanger is steam operated, ensure the steam traps are in operating condition.				
When the heat exchanger is not in use, isolate the energy supply.				

Table 6.10: A shortlist of improvement opportunities for lighting

Housekeeping	Control	Modification	Integration	Alternative
Implement routine cleaning regimes for the light fittings.	Carry out a survey to identify the location of light switches for each light fitting, and lux levels.	Maximize the use of daylight.		
When light fittings are faulty, upgrade to models with high efficiency.	Identify areas where lighting is not necessary, remove these light fittings and/or disable the light switches.	Consider introducing daylight by the use of transparent panels and tubes.		
		Replace tungsten filament lighting with compact fluorescent lighting or LEDs.		
	In areas where the existing lux levels are too high, consider reducing the rating for each light fitting or reducing the number of light fittings.	Replace T12 and T8 light fittings with T5 light fittings or LEDs.		
		Replace tungsten halogen lighting with metal halide lighting or LEDs.		
	For areas with low occupancy, implement a system to turn off the lights, e.g. last out to turn off policy, or the use of passive infrared detectors (PIRs).	Replace floodlights with metal halide lighting or LEDs.		
	For areas where large banks of lighting are	Consider reducing lighting voltage.		

Chapter 6 Using energy audit processes to maximize energy savings

Housekeeping	Control	Modification	Integration	Alternative
	controlled by a very small number of switches. Review timing where lighting is necessary and implement a lighting policy. Consider the use of task-specific lighting.			

Table 6.11: A shortlist of improvement opportunities for road transport

Housekeeping	Control	Modification	Integration	Alternative
Ensure the vehicle does not carry any unnecessary goods and materials.	Utilize low kinematic viscosity engine oil.	Utilize higher efficiency engines.	Review and optimize the number of and the location of transport hubs for the whole supply chain.	
Ensure all maintenance requirements are complete to prescribed standards.	Reduce engine idling. Reduce speeds to match optimal engine fuel consumption.	Utilize vehicles with modern construction materials and lower weight.	Consider sharing transport requirements with other companies with similar routes and/or requirements.	
Energy efficient driver training to anticipate the actions of other drivers and act accordingly.	Utilize automated real time monitoring for driver feedback. Improve scheduling to maximize passenger numbers, capacity loadings and utilization.	Utilize fittings to improve aerodynamics		
Ensure wheels are correctly inflated.	Improve loading sequence and orientation to maximize empty space on the vehicle.			
Ensure good wheel alignment.	Utilize low resistance wheels.			
Drive at constant speeds rather than frequent acceleration and breaking.	Minimize the use of excessive vehicle body paint.			
Move deliveries to off-peak hours to minimize congestion.				

Chapter 6 Using energy audit processes to maximize energy savings

Housekeeping	Control	Modification	Integration	Alternative
	Route planning to minimize stops, starts, and distance travelled.			

Table 6.12: A shortlist of improvement opportunities for aviation

Housekeeping	Control	Modification	Integration	Alternative
Ensure the vehicle does not carry any unnecessary goods and materials. Ensure all maintenance requirements are completed to prescribed standards.	Consider single engine taxiing. Consider the use of reduced thrust take-offs. Consider the use of continuous and constant decent approaches. Utilize low resistance paints. Minimize the need for excessive body paint. Improve scheduling to maximize passenger numbers, capacity loadings and utilization. Consider code-sharing opportunities with other transport providers.	Utilize modern vehicles which utilize more efficient engines, are more aerodynamically efficient and use lighter weight materials.		

Chapter 7   Choosing the right team

As demonstrated in earlier chapters, it is important to be clear on the objective (the reasons for carrying out an energy audit), the required output from the energy audit and the scope (the area, equipment and systems to study). With a defined objective and scope in mind, a team of energy auditors competent for the scope can be formed. Also mentioned in the previous chapters, an organization can choose its energy auditor from in-house resources, external resources or a mixture of both.

In this book, the phrase 'energy audit team' is a team of people an organization puts together to carry out an energy audit or a series of energy audits. The phrase 'energy auditor' refers to the individuals that make up the energy audit team. In instances where there are more than one person in the team, and where an energy audit is carried out for the purpose of ESOS compliance, the energy audit is to be led by a competent person.

Energy auditors, especially when they are sourced externally, think that the organizations hire them because of their cool use of intellect and unsurpassed technical skills and expertise. An energy auditor's skill is purely an initial set of knowledge moulded by a collection of learning and refinement over many years. Many of them spend their working lives (so far) to perfect and achieve, in their minds, technical 'excellence'. According to a survey by KPMG, successful professionals rate their success at 80 per cent on their technical prowess and at 20 per cent on their ability to generate relationships and connect with clients.[109]

Organizations wanting an energy audit need to do so by engaging not only the most competent energy auditor for their needs, but also the most 'user-friendly' auditor to work with the organization and its employees. In reality, unless the managers are themselves expert energy auditors, they are not able to assess the technical capabilities and competencies of their energy auditors. Many organizations place more emphasis on the energy auditor's ability to connect with them. In the same survey by KPMG, the professionals' clients indicated that around 80 per cent of the time, selection is based on people skills, and 20 per cent is based on technical ability.

Therefore, in order to maximize the value from carrying out an energy audit and to maximize the potential for energy reduction, instead of choosing an energy audit team purely on its technical merits – which, according to KPMG's survey, it will likely fail to do – organizations should become aware of their own 'emotional intelligence' and take full advantage of it. This involves balancing five aspects of professional services engagements:

1) hiring preferences;
2) energy audit competence;
3) associated skills;
4) likeability;
5) other 'small print'.

Hiring preferences

Very frequently, when managers in an organization want to identify a suitable energy auditor, they base their decision on either the individual's educational qualifications, previous job titles, links to professional associations and social standing, or how big the energy auditor's company is or how big its fee is. They do so, unknowingly, by assuming that branding and cost are a proxy for measurable results and value that can be gained by working with the energy auditor.

In a survey by IChemE's Consultancy Special Interest Group, more than 60 per cent of managers sourced for external support via personal contacts.[110] (See Figure 7.1). Then, up to 55 per cent of managers narrow down the available pool of external resources by looking for educational qualifications, previous job titles,[111] professional associations and social standing, and/or who they work for. In the context of an energy audit, some common job titles are energy engineer, energy manager, energy auditor and energy consultant.

Alternatively, companies look for energy auditors who are members of professional institutions and/or associations. Membership with a professional institution and/or association does not necessarily make an energy auditor competent in carrying out an energy audit of the specified scope – it is merely an indicator that the energy auditor meets the minimum criteria for membership with the association. Furthermore, if more than one energy auditor has membership with the same professional association, the criteria for selecting and shortlisting based on professional membership will not differentiate between the two.

This way of sourcing energy auditors significantly limits the pool of external resources available to carry out the work. There are many smaller companies and specialist companies in the market that can support an organization. They may be equally qualified and experienced. In the wake of the financial crisis, cost pressures on organizations have

Chapter 7 Choosing the right team

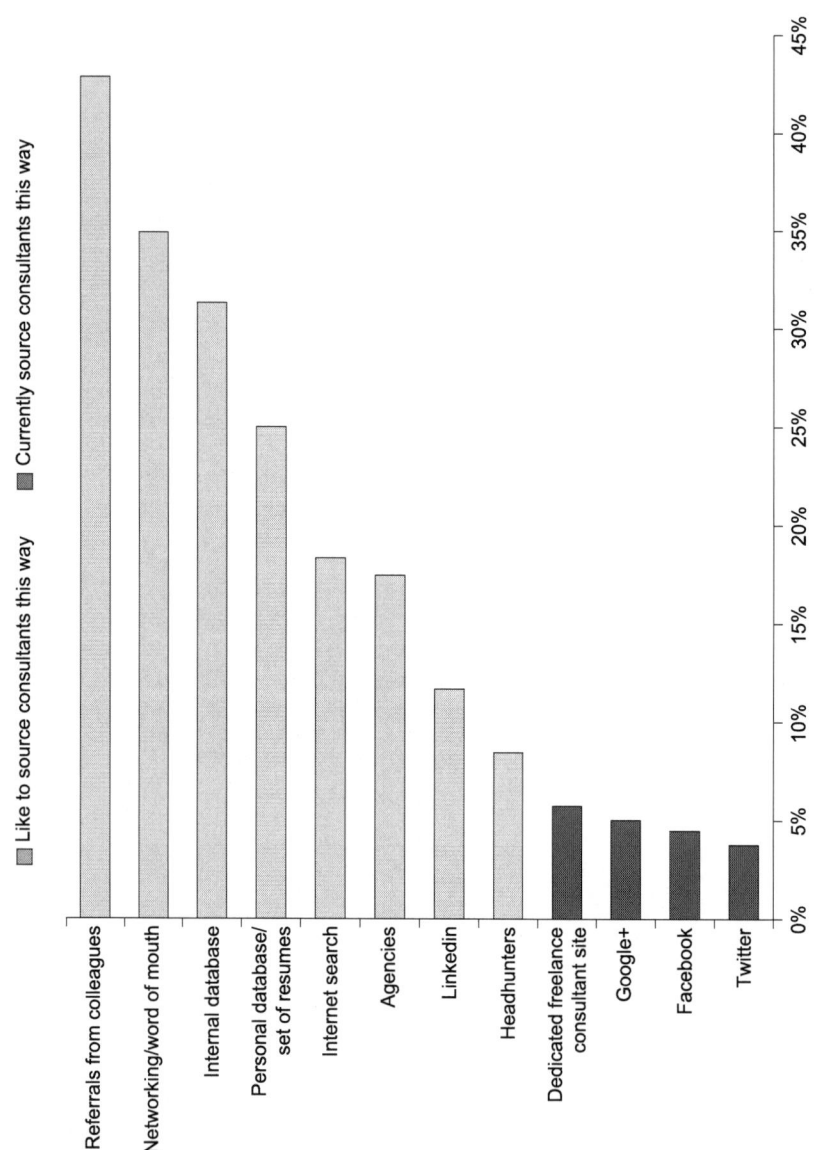

Figure 7.1: Preferred method of sourcing consultancy services.

(Data adapted from: Currie, M. 2013. Communication for Consultants. IChemE Consultancy Special Interest Group webinar)

led many managers to question the bundled services and cost-benefits of buying from a one-stop professional services firm.

In fact, over the last 20 years, there has been a 1,600 per cent increase in experienced professionals going independent or forming small, specialist consultancies.[112] Whatever their reason for going independent, these small-sized companies tend to have a small area of specialism, greater speed and flexibility to respond to the organization's needs and, usually, a lower cost base than a large professional services company. These reasons translate into a cost advantage for the client, as smaller professional services companies do not carry the weight of large, fixed costs, the administrative burden, expensive real estate, and recruitment and training costs.

Legal councils, and management consultancy and technical consultancy industries are beginning to see disruptive service innovations from these smaller rivals, each performing a specific task, making up the whole scope of work required by an organization. Organizations are continuing to want to unbundle what they traditionally bought from one-stop-shop service providers, preferring to pay for the services they need, at greater speeds and responsiveness, and with greater control.[113]

Similar to medical professionals, the skills and expertise of an energy auditor could range from those of a general practitioner to those of a consultant specializing in a specific technology or techniques. Organizations should first determine if they need a general practitioner or a specialist.

In a broad-brush, sweeping manner, a general practitioner has a good command of generic skills across several technologies and techniques. The advantage of a general practitioner is in carrying out a high-level overview of the organization. A wide scope of work allows the energy auditor to break down a silo mentality, adopt a bird's eye view of energy use, and identify and bridge energy reduction opportunities from various parts of the business. Sometimes, energy 'waste' from one part of the organization can be a 'useful' energy source for a different part of the same organization.

A specialist brings specific skills based on specific machines or specific systems that a generalist lacks. The advantage a specialist brings to the organization is their in-depth knowledge of specific machines or processes. It could also be specific technical knowledge that the organization wishes to use, e.g. 'dynamic simulation' or 'thermal integration'. However, in general, though there are exceptions, a specialist tends to lack the breadth a general practitioner brings.

As such, the advantage of choosing one over the other very much depends on the scope of the energy audit. It may be the case that an organization chooses a general practitioner for an initial diagnosis and

Chapter 7 Choosing the right team

investigations, and then chooses a specialist consultant for specific recommendations and/or investigations.

Once the hiring preferences have been established, the organization can then assess the technical aspect of engagement: the competency of the energy auditor. This assessment may also be supplemented by carrying out further background checks, such as reviewing the information on their website, reviewing their résumé, reviewing case studies, visiting or having a conversation with a selection of their previous clients, or obtaining recommendations from word-of-mouth contacts.

Energy audit competence

As in the case of GE and GSK, an organization could use internal company personnel to carry out an energy audit. Typically, they will have a good understanding and knowledge of how the organization operates, historical insight into the plant and equipment performance, and familiarity with how different people interact with existing equipment.

Unless the internal resources have an education in energy calculation, or have been provided with energy management and energy audit training, they tend to be less adept at quantifying the opportunities and require assistance and support. Another area where internal resources need support is in areas away from the business end of the company, i.e. the boiler plants, electrical substations, electricity distribution network, ventilation systems, heating and cooling systems, transport, etc.

An externally sourced energy auditor brings their familiarity of the equipment and the systems, gained from education and/or training, and suitably supplemented by practical, hands-on work. This experience may come from operating, maintaining, designing and commissioning, inspecting, and carrying out similar energy audits of, similar equipment and technology in multiple organizations.

As such, companies seldom carry out energy audits purely on their own. Energy audits are frequently aided by equipment manufacturers' manuals, energy audit guidebooks, checklists, external energy auditors and EPC companies. Together, they form the energy audit team for the organization.

With a defined scope for the energy audit, the organization is able to match the skills and experience of the team with the scope of energy use, technology and systems. For example, a person with skills from auditing a high-pressure steam boiler in a power plant would be less of a match when carrying out the same activity on a domestic or commercial low-pressure, hot water boiler. This is because the technologies, size of equipment, expected norms and practices, and range of improvement possibilities are different. As such, the energy auditor's competence in the power plant is of little value to the scope of work required.

There is a drawback to engaging an energy auditor from an external resource: they do not have the specific knowledge that internal employees bring to the energy audit. Recalling the top 10 reasons why energy management efforts fail in an organization, shown in Figure 2.4, an external energy auditor is also not able to influence the organization's resistance to change, leadership attitudes towards energy investments, and high financial and procedural hurdles, and assess many of the non-energy-related benefits from energy reduction opportunities.

PAS 51215 [114] describes the range of knowledge, skills and experience of an ESOS lead energy assessor. It documents the broad-based competencies necessary to develop technically accurate and insightful energy audits. It acts as a minimum standard with which UK-based energy auditors should compare, and aspire to develop their skills. This minimum set of skills inventory is as follows:

- understanding the operational context of the organization being assessed;
- familiarity with, and the ability to apply, the requirements of energy efficiency assessment methods;
- scoping an energy efficiency assessment, as applicable to the organization being assessed;
- understanding, in detail, the energy use and energy systems applicable to the organization being assessed;
- managing energy efficiency assessment teams and budgets;
- understanding the techniques of measuring, sampling, sub-metering, and establishing an energy balance;
- interpreting data, including analysing and scrutinizing energy use, energy consumption and energy performance data;
- identifying, quantifying, ranking and prioritizing opportunities for improvement;
- managing working relationships;
- preparing and presenting a technical and a non-technical report for an energy efficiency assessment.

PAS 51215 goes further by giving examples of activities that may be used as evidence of skills and experience in relation to each item in the skills inventory. While it was developed specifically for ESOS and is applicable only to lead energy assessors carrying out ESOS-compliant energy assessments for large enterprises, it provides a very good yardstick for identifying and engaging with a competent energy auditor.

There is a specific reason why PAS 51215 describes 'competencies' rather than a 'qualification': a person's competencies are a collection of relevant knowledge, and their application and refinement over time. Attending a training course and passing an examination, also known as obtaining a qualification, only demonstrate that the person has completed a course of study. It does not imply that the person has understood and assimilated the training into knowledge, or is capable of carrying out an energy audit.

The role of lead energy assessor

As mentioned earlier, all ESOS assessment needs to be led by and/or reviewed and approved by a qualified ESOS lead assessor. Within the context of ESOS, the duties of a lead assessor are:

- leading and managing a compliance ESOS energy assessment; or
- reviewing and approving the output of an ESOS assessment carried out by others.

This means that the ESOS lead assessor has a larger role and responsibility than purely leading an energy audit exercise. The lead assessor also has to:

- support the organizations to assess its eligibility to participate in ESOS based on data;
- support the organization to define the scope and boundary of the ESOS assessment;
- assess the energy users that are exempted from carrying out an energy audit;
- ensure that the outcome of all energy audits meet the requirements of ESOS assessment.
    o The lead energy assessor may be the lead energy auditor for a team of energy audit teams or reviews and approves the output of energy audits carried out by other parties.
    o Check and review that the energy consumption profiles exist in the energy audit report and are relevant to the organization.
    o Check and review that opportunities for improvement are based on 12 months of data and that the energy audit is based on appropriate calculations to determine potential savings.
    o Check that practicable and relevant energy reduction opportunities have been identified.
    o Where practicable, the opportunities have been assessed based on life-cycle cost analysis instead of simple payback.

Becoming an ESOS lead energy assessor

To become an ESOS lead energy assessor, energy auditors should apply to become registered with a professional institution and/or association whose professional register has been assessed by the Environment Agency and is satisfied that their competency is in line with PAS 51215. A definitive list of professional registers is published on www.gov.uk/energy-savings-opportunity-scheme-esos. This list of professional institutions and associations is re-approved every four years in line with the compliance dates.

Associated skills

The breadth of the energy auditor's skills needed to carry out an energy audit is only one side of the coin. A spectrum of complementary skills and expertise is necessary for the speedy and timely identification of opportunities, the visualization of the steps of implementation, the budgeting of capital costs and, if necessary, the design and implementation of the agreed opportunities.

PAS 51215 shortlists additional skills that are complementary to an energy auditor's main skill set (reproduced in Table 7.1). They are grouped into technical skills and managerial skills. Mastery of these complementary skills comes with a good foundation in education and practical experience. To a large extent, it is almost always impossible to gain the energy audit skills described in the previous section without first having developed the complementary skills in a specific area and/or technology.

Table 7.1: Complementary technical and managerial skills

Technical skills	Managerial skills
Understanding, and application, of material and energy balance	Communicating the outcome of an energy efficiency assessment
Understanding, and application, of fluid flow	Training
Understanding, and application, of heat transfer	Stakeholder engagement
	Managing change
Understanding, and application, of transport systems	Economic valuation of opportunities for improvement
Understanding, and application, of electrical systems	Generating a business case for implementing identified opportunities for improvement
Measuring, sampling, sub-metering and interpreting results	
Understanding the role of operating procedures and their impact on energy performance	

Chapter 7 Choosing the right team

Technical skills	Managerial skills
Conceptual design, technical and economic evaluation	
Project planning and deployment	
Understanding the importance of maintenance	

(Source: PAS 51215:2014)

The active use of these skills allows the energy auditor to assess the energy use and check that their assumptions are valid, and enables them to minimize the time it takes to define the constructability of the opportunity, the rough capital costs, to engage in intelligent conversations with the whole organization (from the shop floor to the managers and the boardroom), and to find the additional value or benefit that the opportunity can bring to the organization. More importantly, it allows the energy auditor to achieve these points based on facts and science (or engineering) that are real and relevant to the organization rather than using rules of thumb and assumptions.

It is almost always certain that no one person can become competent in all of the skills listed in Table 7.1. Acknowledging this fact is often very difficult for professionals, especially external energy auditors. This difficulty arises from the three innate beliefs that acknowledging this fact makes the external resource look incompetent and feel inferior to the client and, as a result, they will lose potential business. Therefore, external energy auditors tend to shy away from acknowledging their shortcomings, and include only a small pool of 'trusted friends' they know to complement their shortcomings and who will not let them down.

Also as a consequence of this innate fear, many external energy auditors will only introduce and/or recommend trusted friends and co-workers, even if there is a person or company that would be more suited to the task. Mark de Rond quotes a statement from the KPMG study mentioned earlier: 'All too often I find the root cause is the natural instinct to protect what we have and not risk introducing a colleague who may put all of our achievements at risk.'[115]

Organizations are turning around and beginning to value what Patrick Lencioni, a New York bestselling author, calls 'Getting Naked' [116], energy auditors that:

- are more concerned about the organization doing the right thing than doing the easy thing;

Associated skills

- ask and report the difficult but truthful questions rather than prioritize saving face; and
- will acknowledge their own limitations and willingness to learn.

Do, also, assess energy auditors on these additional and complementary skills.

Also included in PAS 51215, and shown in Table 7.1, is a range of non-technical skills: the soft people and managerial skills. Many technically minded professionals will also shy away from these soft skills. They may see them as menial, unchallenging and not important. The Environmental Defense Fund (EDF) Climate Corps programme would beg to differ on this viewpoint.

EDF Climate Corps is an annual programme where a group of postgraduate students in a wide range of studies (business administration, management, science, the arts and engineering) are trained in basic energy auditing and deployed as summer interns over a 12-week period. During these 12 weeks, an EDF fellow supports organizations to identify opportunities for improving energy efficiency, delivering the energy savings and cutting GHG emissions.

Now in its sixth year (2013), EDF Climate Corps fellows have helped to save nearly £1.3 billion in energy savings at 106 organizations, including those of Apple, Facebook, adidas, General Motors, Verizon, the US Army, Chicago Public Schools and many more. In particular, EDF Climate Corps noted a number of organizations providing feedback on how beneficial it was to have fellows that were:[117]

- able to communicate effectively across the organization, moving information from the shop floor and engineering (who know about energy efficiencies and how to implement them) up to the chief financial officers (CFOs) and executives, using terminologies and words that resonated with each stakeholder;
- able to moderate and arbitrate between several stakeholders and come up with alternative and innovative win–win solutions in areas involving 'split incentives', i.e. where there is an interface between tenants and property owners. Some examples are:
  o having a split scope of responsibilities that makes it difficult to justify long-term paybacks independent of the other party;
  o financing energy savings between multiple parties; and
  o opportunities that involve other non-energy-related savings, such as improved worker productivity from a better work environment.
- able to identify and design employee engagement programmes and behaviour change strategies as a key strategy to reducing energy consumption.

Energy Audits

Likeability

In addition to identifying and quantifying opportunities for energy reduction, managers frequently rely on consultants to support them to develop the capabilities of an internal change team, to facilitate the change, and to act as a sounding board on other energy reduction programmes. As suggested earlier, energy auditors are also used to facilitate and enhance interaction within the business.

Therefore, apart from possessing the breadth and depth of skills needed for an energy audit, and a range of complementary skills, there are other reasons why energy auditors are engaged by an organization.

The degree to which the external resource fits in with the organization's ways of working and culture is important. In order for the business to gain maximum value from engaging the energy auditor, both the managers and the employees within the business need to be able to communicate with the consultant in a manner that is both fulfilling and meets the business goals.

An energy auditor could be the best from the available pool of resources. The same energy auditor could be very difficult to work with. For example, they talk in very technically based jargon that nobody understands and will not change their use of jargon. Alternatively, they communicate in a manner that does not inspire or instil confidence, e.g. they are insulting, they are perceived to be too aggressive, they belittle the managers who know least, or they talk about opportunities in a 'black box' manner.

An energy auditor that communicates 'black box' may introduce an unknown risk and uncertainty, which cannot be assessed prior to it being implemented. In the eyes of the organization, this could be deemed risky and unacceptable.

Many skills (and, sometimes, trade-offs) are required from both parties. These range from the active participation of the managers and engineers of the business to listening to and discussing ideas, to authentic communication and a commitment to building a good relationship.

Listen to the presentation and/or sales pitch of the energy auditor. Are they focusing the talk on their 'skills, superiority and experience' or on how they could 'work with the organization to deliver XYZ'? How did they engage with their previous customers? Does the energy auditor instil confidence in you about their capability? Is their track record well received and gained in a similar industry or cluster? How do they measure a successful business relationship? Do they base their success on 'Mr CEO says he or she is satisfied' or 'We did this and that, and achieved ABC results'? Do ask the energy auditor: 'How would you approach the issue at hand?' or 'What obstacles do you foresee and how do you recommend we overcome these issues?'

Through these conversations, observe and make an objective assessment on whether the managers and employees of the organization can work with the energy auditor, whether the energy auditor 'feels right' for the needs of the company. Does the energy auditor come across as authentic or obsequious? Sometimes, managers choose to work with the energy auditor who is congenial and likeable in preference to the most competent one.[118] As long as the decision to engage with an energy auditor is made consciously, this is perfectly acceptable, too.

Other 'small print'

Objectivity (or independence), confidentiality and professional indemnity insurance are some of the most common issues managers of a company and professional associations focus on when entering into a commercial engagement.

Objectivity

The objectivity of an energy audit refers to it being carried out in an independent and impartial manner. This means that the energy audit team carries out an energy audit of the agreed scope without preconceived solutions in mind. In theory, this allows the energy audit team to match the improvement opportunity with the solution that fits best, thus giving the organization the freedom to purchase the necessary equipment it deems as valuable and economic, for example, from a range of suppliers.

In practice, this is slightly more complicated to enforce. The following list gives some examples where objectivity could be in shades of grey rather than in black and white:

- the energy auditor is carrying out an energy audit where a close friend or colleague is responsible for the area. Is the energy auditor going to be cutting their close friend or colleague some slack and painting a better than average picture of energy performance? *Or* is the energy auditor going to report it as they find it, regardless of their social standing with that friend?
- the energy auditor is carrying out an energy audit of an area managed by a sworn enemy. Is the energy auditor going to be more critical of the energy audit? *Or* is the energy auditor going to report it as they find it, regardless of their social standing with that person?
- during the energy audit, the energy auditor refers to publicly available technical brochures from a well-known equipment

Chapter 7 Choosing the right team

manufacturer. Is the energy auditor colluding with the said manufacturer? *Or* is the energy auditor purely using publicly available information to identify solutions that fit best?
- a well-known equipment manufacturer carries out an energy audit on a piece of equipment or a system that is also a product of the equipment manufacturer. Is the energy auditor going to identify all possible opportunities for improvement, using all available technology? *Or* is the energy auditor going to limit their findings to those products that their company sells? Will they recommend replacing a fully functional piece of equipment with one manufactured by their company?
- the energy auditor has several friends who provide various kinds of energy-efficient technologies. Is the energy auditor going to recommend the technologies and services of their friends? *Or* is the energy auditor going to recommend the technologies and services that best fit the client's needs?
- the energy auditor also operates a project management company. Is the energy auditor going to recommend only projects that their own company can deliver? *Or* is the energy auditor going to recommend all identified opportunities?
- an electricity supplier, gas supplier or ESCO carries out an energy audit for the company. There is no cost to the client. The supplier recoups the cost of the energy audit, and investment costs, from the savings generated from the project. Is the energy auditor going to maximize the capital costs for the supplier? *Or* is the energy auditor going to maximize the energy-saving potential for the client?

As demonstrated above, in the real world, it is very difficult to enforce a strict objectivity rule. Companies should be aware and be vigilant when choosing the energy auditors for the company, and during the energy audit process.

A good way to overcome this issue is to review many case studies, sample reports and previous projects from the energy auditor. A consistent pattern should emerge.

Even with the best of intentions, it may not be possible to remove fully any bias. In this instance, the best solution may be for the organization to:

- require the energy auditor to disclose their commercial and technical interests before engaging them; and
- have a clear and defined end goal/exit strategy in mind before engaging with the energy auditor, i.e. a defined scope and tasks.

Confidentiality

Protecting the information of the business, and information about the engagement with the energy auditor, is normally the second topic on which organizations focus. This is even more the case if there are commercially sensitive or trade secrets, and/or licensed processes, involved in the scope of the energy audit. An example could be that the energy audit will involve key technologies or techniques employed by the company to differentiate it from its competitors. Other examples are a military equipment manufacturer, a financial services provider or a central government building.

The confidentiality of energy auditors is exercised by non-disclosure agreements between the energy auditor and the organization involved. There are several levels of confidentiality clauses. Managers should consider which level of confidentiality is suitable and appropriate for the work that is to be carried out. These are:

- entry-level: commercial information about the engagement, and information pertaining to specific processes and/or licensed processes, is not disclosed;
- all information: no information collected and used is disclosed. This includes the details of the energy audit and any opportunities found;
- time-bound: large, multinational companies with strong buying power, or companies handling sensitive information, sometimes embargo all information for a specified time frame.

As a knee-jerk reaction, many companies will apply the maximum level of confidentiality requirements. While it can give an organization peace of mind exercising lots of confidentiality clauses when engaging an energy auditor, the same principle can also limit the scope of the energy audit, and opportunities to achieve maximum energy reduction.

Applying large restrictions on confidentiality may also limit the range of consultants available and increase the costs of the engagement. It is advisable to make sure that suitable clauses commensurate with the scope and expectations are chosen and that they are discussed with the energy auditor.

Professional indemnity insurance

All externally sourced energy auditors have professional indemnity insurance and have an allocated working capital. Some organizations require these external energy auditors to have a minimum amount of professional indemnity, or liability, insurance or a specified amount of paid-up capital, before authorizing them as an authorized energy auditor. Requiring a minimum amount of professional indemnity insurance or a specified amount of paid-up capital for professional services is an artificial comfort for organizations.

In reality, the applicability of such a requirement does little to help the energy auditors minimize mistakes. Working jointly with, and alongside, the energy auditors to understand the collected data helps to minimize mistakes and errors. This way of working increases the quality of the working relationship and generates value from the commercial engagement.

Buyers beware

Some managers presume that by engaging multiple energy auditors or, perhaps, pre-qualifying the energy auditors using a type 1 audit, all of the energy auditors will come to the same conclusions in terms of energy reduction and capital costs. If the organization expects several energy auditors to use the same energy audit standard, have access to the same data, and have the same complementary competences and skills, to arrive at the exact same recommendations, it will be disappointed.

The best example of this comes from the US Department of Energy's Energy Efficient Buildings (EEB) Hub[11] – Navy Yard Building 101, Philadelphia.[119] This building, originally built in 1911, provides 61,700 square feet of office space over three wings. The plant rooms are located in the basement and in the attic, which is also utilized as a storeroom.

As a research centre for energy efficiency, this building uses 509 sensor points to collect 1,048 pieces of energy-related data at one minute intervals. These are key pieces of information pertaining to indoor air quality [CO_2, carbon monoxide (CO), total volatile organic compounds, small particles, indoor temperature and relative humidity], occupant comfort (workstation activities, thermal quality, visual/lighting quality and acoustics) and building energy use (electricity consumption, natural gas consumption, air flows, position of outside air dampers, air leakage and exhaust air).

Three companies were chosen to carry out an energy audit of the building and EEB Hub was expecting the results from the three consultants to be fairly consistent and similar. The actual recommendations and outcomes were far from consistent.

At a cost of $500,000, Consultant A identified nine opportunities to save $60,200 per year, or a 38 per cent energy reduction. Consultant B identified 12 opportunities to save $22,495 per year or 14.5 per cent, with an investment of $150,000. Lastly, Consultant C proposed to

---

[11] Energy Efficient Building Hub is now called Consortium for Building Energy Innovation (CBEI).

implement nine opportunities at a capital cost of $160,000, with the benefit of a 24 per cent reduction or $34,000 per year. When the recommendations were analysed by EEB Hub, only three opportunities were common across Consultants A, B and C's recommendations. Furthermore, the way in which the consultants prioritized the common opportunities was reported to be not in the same order.

As the example demonstrates, a company may engage the best and most co-operative energy auditor in the world. The organization may also insist on all of the items listed in this chapter being followed. There will always be variability in energy audit teams' findings and recommendations. This has little to do with the competencies, experience and skills of the energy audit team. Apart from the items discussed in this chapter, the variability can also arise from the things that the team sees, and from the information that is presented, as well as from the way in which the team chooses to bundle the projects.

It remains a hypothesis that the organization and the energy auditors could ever remove 100 per cent of the biases. The only way to prove or disprove this is to carry out another 'EEB Hub'-styled' experiment using identical triplets with three identical sets of knowledge and career development paths, using identical tools and techniques, and carrying out an energy audit at the same time!

Chapter 8 Implementing energy reduction

Steven Fawkes

Although the large potential for improving energy performance, and the benefits of doing so, are increasingly recognized at global, national, regional and corporate levels, in many situations it remains just that – potential. In order to exploit improved energy performance and enjoy the associated financial and other benefits, organizations and individuals have to actually implement energy reduction actions and projects. Simply carrying out an energy audit, or identifying opportunities, does not bring the benefits, only implementing the measures can do that.

Over the years, improving energy preformance has been beset by governments and managers carrying out energy audits and then putting them on the shelf, without any implementation. Even the EED and ESOS only require companies to identify energy-saving opportunities – not to implement them. Any agents of change must maintain a laser focus on implementation. This chapter looks at some of the key issues around implementation.

It is about management not technology

It is important to start by stressing that the process of improving energy performance is not a technical problem but rather a management problem. Many well-proven technologies can radically improve the energy performance of almost all (if not all!) fuel and electricity use in buildings, industry and transport. In fact, even if no further technologies were developed and we just applied all the existing technologies that are economically viable right now, we would close the 'energy performance gap' and achieve far higher levels of energy savings than we do today. Figure 2.4 shows the range of managerial issues at play.

There are a number of enabling conditions that need to be in place to support a successful energy management programme, but the most important is full support from the leadership of the organization. Even if energy managers or engineers are actively implementing energy reduction projects, they cannot achieve the maximum results without the full support of senior management. Only when senior management is convinced of the need to reduce energy usage, puts in place the appropriate organization and systems, gives it appropriate resources, and insists on the appropriate reporting tools, can an organization achieve its full potential. The evidence from companies such as 3M, The Dow Chemical Company, Sainsbury's and Walmart is that when leadership on energy efficiency is strong, organizations can make significant energy savings year after year, and even decade after decade.

The first challenge for reducing energy consumption is how to truly engage top management and persuade it of the importance of energy efficiency. To do this it is important to talk about the energy performance improvement opportunities in the right language and, in the case of private sector organizations, this will most likely be about profit and productivity. A commonly heard comment on energy performance in many organizations with low or medium energy intensity is that energy costs are only a small fraction of total costs and therefore it is not worth taking action. Potential energy savings should be compared to profit rather than to total costs and, in many cases, achievable energy savings will provide a significant proportion of profit. Given that energy savings go straight to the bottom line, the additional profit available, and the relative ease of achieving it from energy efficiency, should be compared to the increase in revenue that would be necessary to achieve the same increase in profit, and the difficulties of doing that, particularly in the current economic climate. Energy costs should always be thought of as a controllable cost, far more controllable than some other costs, such as labour.

Responsibility and accountability for energy use has to be given to those who can actually control energy consumption and this means building/facility/profit centre managers at an appropriate level. In many organizations, responsibility for reducing energy use is given to an 'energy manager', but this is a mistake; an 'energy manager' cannot control the activities within profit centres that lead to energy being used. The energy manager or energy specialist can only ever be a service provider, an internal consultant and provider of information and advice. The best results can only be achieved by giving explicit responsibility and accountability to profit centre managers and, in particular, by adding energy performance improvement to managers' targets and incentive programmes. The flip side of giving responsibility for energy use to a manager is that the manager must be able to control the variables that affect energy use, and they must also have access to timely and accurate information on energy use, as well as appropriate benchmarks. Giving the responsibility without both the ability to act and the use of energy performance information, just like giving responsibility to the wrong person, cannot work.

Responsibility for energy performance should be passed as far down the organization, and as close to the point of energy use, as possible – doing

Chapter 8 Implementing energy reduction

this helps to engage employees and can bring out their responsibility for, and inspire creative new ways of, reducing energy use. How far devolved responsibility can be taken will depend on the corporate culture, and the ability to provide regular information from sub-metering.

In order to mount an effective energy management programme, an organization needs an effective energy management system, such as one based on ISO 50001 – here we are talking about a human system rather than an electronic or software system, although, of course, it may well use advanced electronics or software in the form of a management information system and sophisticated control technologies. The purposes of an energy management system should be to:

- enable the management of energy consumption and costs day by day, hour by hour and, even, minute by minute, by identifying waste and facilitating prompt and effective action to stop the waste;
- identify, evaluate and implement energy performance improvement opportunities (including both 'soft' projects, such as motivation schemes, as well as 'hard' capital projects).

A good energy management system will change the questions being asked about energy from the traditional questions such as 'How can I get the energy supply capacity I need?' to 'How much energy do I need for that application?' and 'What is the optimum investment in fuel, conversion and distribution mix to supply it?' Asking different questions about energy leads to different answers, radically different in many cases. When a good energy management system is applied properly the results can be:

- optimized energy use in buildings, processes and transport;
- fuel-efficient supply processes;
- new management models;
- dramatically reduced emissions;
- enhanced competitiveness from:
    o reduced operating costs;
    o improved environmental performance and credits;
    o increased security of energy supply.

Project identification, development and evaluation

To identify or create investment opportunities requires internal information about energy costs to be creatively synthesized with inside and outside knowledge of appropriate technical concepts. Creativity can always redefine what is appropriate or possible. When considering technical measures, an explicit decision on the appropriate level of innovation or technical risk is needed. Some organizations may be willing to take on the risk of using a prototype or early version of a new piece of technology, but, for most organizations, the simple adoption of existing technologies is the more appropriate response for energy performance improvement investments – usually, there is no desire or need to take risks with technology.

Once the improvement opportunities are identified, they must be appraised technically, contextually and economically. These appraisals should interact in an iterative process that itself can lead to new ideas. This is essentially a process of project development.

Technical appraisal

Technical appraisal covers the applicability and suitability of the improvement opportunity to the application. This should, as far as possible, be based on a representative set of operational and other data described in Chapter 5. For example, the suitability of downsizing the motor of a fan versus the application of a VSD or finding options to maximize the energy reduction for the energy use.

Failure to undertake adequate technical appraisal often leads to wrong or sub-optimal application of the energy improvement technique. An example would be application of VSD where a smaller motor is necessary or installing a condensing boiler when the application does not allow the boiler flue gases to condense. Such failure would also lead to a larger capital investment than necessary.

Contextual appraisal

Energy performance improvement projects can, and, in the real world, often do, interact with each other, as well as interacting with process or operational changes. These interactions can be either positively or negatively synergistic. Assembling a portfolio of potential projects, maintaining them in a database, and regularly re-evaluating them independently, as well as together, is a useful technique.

Contextual appraisal covers interactions with other management decisions being considered or made, such as the possible closure of a facility or a relocation, and other management processes, such as procurement, sustainability, property portfolio management and supply chain management. Failure to undertake adequate contextual appraisal often leads to wasted investment. An example would be where recently retrofitted equipment has to be removed during a relocation, or due to a planned closure of a plant, which had not been properly considered. Project development also has to take into account maintenance and operational factors, as well as environmental factors such as noise, vibrations and emissions.

Chapter 8 Implementing energy reduction

Economic evaluation

As mentioned earlier, the main motivation for improving energy performance is economic – cost reduction. Although there was a period when many believed it was more about reducing carbon or CO_2 emissions, for most organizations, most of the time, energy performance improvement is about saving money – becoming more productive in the use of resources, and in reducing wastage. Improved energy performance does, however, bring with it a number of other benefits, some of which can have an additional economic benefit that needs to be evaluated and recognized in any investment decision. These include:

- reduced exposure to energy price volatility;
- a reduced need to invest in new energy supply capacity, e.g. larger electrical connections;
- the improved health, well-being and productivity of employees;
- an improved quality of production;
- reduced operation and maintenance (O&M) costs;
- a reduced threat of energy supply disruptions;
- increased property values;
- reduced emissions of CO_2 (with associated savings through the reduced impact of carbon taxes or trading schemes such as the CRC and EU ETS);
- reduced emissions of other pollutants, which may have an impact on the freedom to operate production plants.

Also mentioned in Chapter 1, outside the boundaries of the energy user and/or organization, at a national level, improving energy performance can also bring an economic benefit through:

- a reduced need to invest in energy supply infrastructure;
- a reduced cost of importing fuel/electricity;
- reduced costs of emissions/air pollution;
- job creation.

It is important when evaluating the economic benefits of energy performance improvement opportunities that all the benefits (and costs) – or at least those within the organization's boundaries – be properly included in the analysis. Very often the focus is just on the energy savings and the other economic benefits are ignored.

Much has been written about how to assess the economics of energy performance improvement investments. At the end of the day they should be assessed in the same way as every other investment opportunity facing the organization or third-party investor. For most commercial organizations, economic investment criteria will be assessed by a DCF technique – either IRR or NPV. DCF techniques are superior to the often used metric of simple payback as the latter does not take into account the magnitude and timing of all project cash flows, nor does it consider how profitable a project will be, only how quickly the investment will be recovered. DCF techniques are also more consistent with a company's objective of maximizing shareholder value. The critical issue is that the energy performance improvement opportunity should be evaluated and presented in a way that meets the investor's needs, whether the investor is internal or external to the firm, and energy specialists need to learn what the preferences, needs, systems and processes are, in order to navigate them successfully. Very often, energy specialists fail to do this and experience frustration when bidding for resources to get work done. The first essential task for anyone seeking to secure financing for energy performance improvement projects is to understand the investment criteria and assessment methods of the investor, be they internal or external.

Reducing uncertainty

A major issue in the funding of energy performance improvement projects, either internally or externally, is uncertainty over the assessment. The perception among non-technical personnel (such as management, accounts, finance directors and other disciplines) is that the savings numbers being presented have a high degree of uncertainty – and often they do. This is caused by several factors:

- the calculation methods;
- the base assumptions about characteristics of the existing system [e.g. just exactly what is the heat transfer coefficient value (or U value) of that wall you are considering insulating? Is the quoted figure right of realistic? In reality it may be very different to what the tables say it should be];
- existing environmental conditions;
- operating hours; and
- user behaviour patterns.

Managers in the organization need to work to systematically reduce the uncertainty of their investment case. This can be done by better numerical and statistical analysis including:

- identifying critical assumptions;
- carrying out a sensitivity analysis to determine the effect of a variation on any of the assumptions;
- assigning probabilities to the variations;
- working to reduce the uncertainty of those variables that will have the biggest impact on investment returns;
- producing a risk-adjusted return.

Investor uncertainty can also be reduced by the application of standard protocols, such as those developed in the USA by the Investor Confidence Project (www.eeperformance.org), and standardized measurement and verification (M&V) tools, such as ISO 50015 *Energy management systems – Measurement and verification of energy performance of organization – General principles and guidance*[12] and the International Performance Measurement and Verification Protocol, IPMVP (www.evo.org).

Financing energy performance improvement investments

The funds for energy performance improvement projects can come from either internal funds or external funds. Internal funds are always subject to competitive pressure and energy savings is often categorized as a lower-priority, defensive, cost-cutting investment, which is likely to lose out to offensive spending on new products, capacity increase or marketing. Internal funds are often subject to tight investment criteria, most often expressed as paybacks of between one and three years, which may equate to an IRR of between 90 per cent and 20 per cent for a five-year lifetime project. Experience has shown that for many (if not most) organizations, allocating a capital budget of about 10 per cent of energy spend to energy performance improvement investments for a three- to five-year period is viable without 'running out' of projects.

External funds dedicated to energy performance improvement can have lower investment criteria than internal funds and so the use of such funds, through some sort of outsourcing arrangement or third-party financing, offers the possibility of investing in longer payback projects, such as deep building retrofits. The issues that need to be considered when looking at external funds include:

- What is the nature of the associated outsourcing contract? The energy performance contract has long been the dominant contract form, and is usually the only form discussed, but new types are emerging, such as the efficiency services agreement (ESA), the Managed Energy Services Agreement (MESA®) and the measured energy efficiency transaction structure (MEETS) among other approaches.
- Who is the investor counterparty? Are there credit or reputational risks?
- What is the real price of the capital? Often, in outsourcing deals, the real price of money is obscured by the 'bundling' of services and finance – deals need to be 'unbundled'.
- What is the length of the funding period? What are the penalties for early repayment?
- Who takes which risks associated with the project(s)? Risks include construction, performance and maintenance.

---

[12] At the time of writing, this is in development.

- On whose balance sheet will the financing sit? Off-balance sheet financing, which has long been popular, is becoming more difficult as the accounting rules are being tightened and harmonized across jurisdictions.

Project management

Once a portfolio of energy performance improvement projects has been developed and financing has been secured, the projects then need implementing, which, as with all projects, requires sound project management. The difference between energy performance and other areas is that, typically, an organization is looking at multiple, relatively small, projects, rather than one large project. Project management techniques need to recognize this difference and be appropriate to the size and number of projects. We are seeing the emergence of integrated software tools that combine diagnostics, monitoring and targeting (M&T), M&V, project identification, project management and post-investment assessment.

Post-project analysis

Analysing the results of energy performance improvement investments is sometimes neglected, but is important. First of all, any underperformance can be detected and, hopefully, corrected. Secondly, it is important to learn from experience in order to improve future projects and programmes. In the case of building retrofits, there is often a large gap between predicted savings and actual savings, and post-investment assessment can help to understand and reduce this modelling gap. Sometimes the gap is caused by end user behaviour, which was not considered by a purely technical analysis. Finally, of course, success, when properly measured and communicated, can help to unlock management support and additional funding.

The massive potential for improved energy performance in most, if not all, areas of the economy is a massive energy resource, analogous to an oilfield or a shale gas field – and that resource can almost certainly become our cheapest, cleanest and fastest source of energy. Turning the resource into a source of energy requires the implementation of energy efficiency projects through the application of sound energy management processes and systems, and the tools to do this are well proven but still need to be more widely applied.

We know that when this is done, organizations of all types will be able to profit from exploiting the scarce energy sources for many years, and even decades, to come. Leading examples that can be found in many of

Chapter 8   Implementing energy reduction

the examples in this book and in the media include Dow Chemical,[120] Walmart,[121] Sainsbury's,[122] Telstra [123] and Stena.[124]

Dow Chemical, a heavy user of energy in intensive chemical manufacturing processes, has, since 1990, reduced its energy intensity by 40 per cent, and saved $24 billion and 5,200 trillion Btu, roughly equivalent to the energy use of 48 million US individual family homes.

Sainsbury's has, for many years, implemented energy efficiency projects across its portfolio with great effect. It has recently switched to an absolute emissions reduction target from a relative one, which will increase the pressure to identify and implement energy efficiency and renewable energy projects.

If more organizations could match the impressive level of success in actually implementing energy efficiency achieved by these leading organizations, we would go a long way to resolving many of our energy supply and energy security issues, as well as greatly improving profitability, and reducing the environmental impact of energy use.

Appendix A People aspects/behavioural change

John Mulholland

Every organization employs people. Every organization uses energy. So the interface between the two has to be addressed by any energy management strategy. In fact, people solutions are the focus of 4 of the 13 energy clauses in ISO 50001. And yet, most energy management and energy audit efforts focus on the technical measures required to achieve energy reductions. However, these will not occur unless someone takes action. People are intrinsically linked to energy management.

A well-planned and properly resourced behaviour change programme:

- will deliver cost savings of many times its implementation cost;
- will produce other benefits, such as improved job satisfaction and process improvements;
- can often be delivered within the existing resources over the long term, but short-term initial support may be required.

Furthermore, there is a growing realization that such programmes have a better return on investment than many energy-saving capital projects. Behaviour change programmes also reduce CO_2 emissions more cost-effectively than most other measures.

This appendix examines how to make savings by harnessing employees so that people solutions can be treated as another energy-saving measure, and can be an integrated part of the energy management strategy.

In many organizations an investment of 1 per cent of the total annual energy expenditure can yield savings of 5 per cent or more. Some organizations report savings of up to 20 per cent. Clearly, the potential for savings will depend on the size and nature of the organization, and its starting point.

If an organization has an annual energy expenditure of £1 million and it invests 1 per cent of this (£10,000) in an energy behaviour campaign, the savings can be in the region of £50,000 or more. A key question concerns the amount of energy directly controlled by employees, when assessing potential and setting reduction targets.

Appendix A People aspects/behavioural change

Identifying the current situation

The first step is to examine who has control over energy use in the organization. Everyone is an energy user but not everyone has direct control over its use. For example, in large retail stores, the sales assistants often have no control over energy use in the sales area because BMSs usually control the heating, ventilation and air conditioning (HVAC), and lighting. Therefore, any engagement strategy needs to address those who control the BMS or the limited number of staff who have some manual control. Some of these people may include contractors.

Quite often in industry the Pareto principle applies, where 80 per cent of the energy is controlled by 20 per cent of the employees. The 20 per cent are often production managers and/or senior plant operators, whose decisions can significantly affect energy consumption. In this situation, the 20 per cent need focused attention and job-related energy training that is specific and tailored to identified needs. Most of the effort must go here if significant savings are to be achieved. The remaining 80 per cent of employees might be subject to a communications focus.

In some organizations, such as those in higher education and health care, there are a relatively high number of employees with large, overall energy consumption. Here, the Pareto principle does not apply, because energy use is more dispersed and, therefore, it is important to engage all employees rather than to focus on a few. Sometimes the use of energy representatives or champions by department or building can help to bring focus.

To have an effective plan it is important to understand the interface between people and energy. The culture of the organization must be assessed, along with the management structure, style and communication methods. Different approaches are required for different organizations. Different approaches are also required for different people within organizations.

Employee environmental awareness surveys

A properly designed staff environmental survey can quickly provide information needed to help design an effective campaign. A survey:

- quickly identifies and recruits volunteer energy representatives or champions;
- quantitatively measures staff awareness/motivation levels;
- analyses different types of staff by site, job function and department. This information can be displayed in a matrix so that comparisons can be made;
- can be repeated at a later date, so that shifts in awareness/motivation can be plotted;

- helps to tailor a campaign strategy based on facts;
- identifies 'quick win' opportunities;
- identifies barriers to improved energy performance;
- has an awareness-raising effect.

The questions in a survey need to be clear, concise and easily understood, with the majority provided with multiple-choice answers for quick analysis. Some questions can be open-ended, but these should be restricted in number. Otherwise there will be a large volume of information to read, categorize and process. The questions usually fall into four categories, as follows.

1. Awareness/knowledge

These questions test the respondents' understanding of an energy issue. For example:

> Is the following statement true or false?
>
> 'It is cheaper to leave fluorescent lights turned on than to switch them on and off when needed.'
>
> ☐ True ☐ False ☐ Don't know

2. Motivation

These questions test the respondents' motivation and general attitude to sustainability. For example:

> On a scale of 1 (low) to 10 (high) how interested are you in saving energy at work?

3. Opinion

These questions test the respondents' opinions or perceptions of, or views on, an issue. For example:

> How seriously do you feel our organization takes sustainability?
>
> ☐ Very seriously ☐ Seriously ☐ Not seriously compared ☐ Don't know
>
> to similar organizations

4. Factual

These questions ask the respondent about the respondent – their site, department, job function and gender – to ensure the survey covers a representative sample.

The survey should be short otherwise respondents may lose interest, skip questions or even fail to complete the survey. But if it is too short, the quality and quantity of information provided will be limited. Surveys that take seven to nine minutes to complete are about right.

There is usually a better response if the survey is anonymous. However, respondents will have to leave their details if they wish to enter a prize draw or wish to volunteer to be an energy champion.

A 10 per cent response rate is typical for voluntary surveys. Some organizations report response rates of 50 per cent where the surveys are well promoted face to face to managers, who, in turn, encourage staff to participate. However, a 10 per cent response rate is usually sufficient to get a representative sample. If areas are unrepresented, these can be targeted while the survey is still open, so careful monitoring is required.

The full survey results should be analysed and written up in a report, providing key conclusions and recommendations. The results of the survey should inform the campaign strategy plan and the communications strategy. The full report should be available to all staff on request, but a two-page summary should be widely distributed to all staff.

Campaign strategy plan

At the heart of energy efficiency behaviour change and employee engagement is a clear and concise campaign strategy plan. This plan provides the background, benefits, targets, strategy, methods, responsibilities, activities and timetable. More importantly, the plan:

- describes key messages and target audiences;
- clarifies the organizational engagement objectives;

Campaign strategy plan

- lists the desired outcomes from the engagement;
- lists the channels and means of communication;
- identifies weaknesses, gaps and further opportunities;
- helps to gain support from senior/middle management.

Within the campaign strategy plan should be a clear communications plan identifying key target audiences and the means of communication.

Forming a campaign team

It is important that the programme is not built on a single individual. Otherwise, if that person gets promoted or leaves the company, the initiative can falter. Also, there needs to be a division of responsibility so that the load is shared by a team of committed individuals. In selecting the team, it is important to look for individual contributions so that the team is balanced in terms of capabilities and skills. Often, organizational and interpersonal skills are more important than technical ability.

A typical team might consist of the following:

- campaign manager;
- energy/environmental manager;
- communications/HR manager;
- operations manager;
- maintenance manager;
- external specialist consultant.

Quite often, other key people can be co-opted onto the team for specific reasons. Typically, these might include:

- senior management representative;
- IT manager;
- production managers/supervisors;
- management systems specialist;
- security manager;
- cleaning supervisor;
- trade(s) union representative(s);
- facilities management manager;
- training manager;
- catering manager;
- health and safety manager;
- quality manager.

Each individual will have a different perspective and insights, which can stimulate new ways of thinking and identify opportunities.

Energy Audits

Appendix A People aspects/behavioural change

Suggested structure for a campaign strategy plan

A good structure for a campaign strategy plan is as follows.

- *Introduction:* a brief overview of the campaign. Why a communications strategy is needed. What behaviour or attitudes does the organization want to influence through the strategy, to achieve the cultural norm to think and act sustainably in relation to energy, water, waste, transport and procurement?
- *Background and content:* this aspect describes the campaign aim and whether it has a name, strapline, slogan and/or logo. It explains the organization's annual energy expenditure and CO_2 emissions. It provides details on target reductions over a given period. It covers external drivers, such as rising energy costs, legislation, corporate/social responsibility and the corporate image. It also covers internal drivers, such as budgets, policy, compliance and risk. It describes communications from previous energy reduction programmes. It indicates how strengths will be built upon and weaknesses minimized. It outlines any cultural barriers and how these might be addressed and turned into opportunities.
- *Overall approach:*
  - *message:* the key messages for the campaign. Are they concise and clear so people can act? How can energy-efficient behaviour become mainstream? Who are the energy champions? How do the campaign messages link to current policies, strategic objectives and targets?
  - *market:* who are your target audiences? Are the same messages to be used for all, or are specific messages to be used for particular groups? Are they purely internal communications, or do you intend communicating externally, as well? Might external audiences include contractors, customers, suppliers, local communities, government bodies and the general public? What actions do you want target audiences to take as a result of your communications? For example:
    - promote/endorse the campaign;
    - participate by adapting behaviour/routines/procedures;
    - innovate and identify new opportunities;
    - volunteer in a specific role, such as energy champion;
    - provide management support, such as releasing staff for training or for conducting walkabouts;
  - *means:* list the means or channels of communication currently available. For example: electronic, hard copy and/or face to face;
  - *identify new mechanisms:* for example, provide visual prompts around the organization, such as plasma screens and banners. Assess their effectiveness by consulting with employees. What are considered to be informal means of communication? What is the appropriate timing, frequency and seasonality?

- *Review/revise campaign strategy:* create feedback loops from the staff on the effectiveness of the communications. Review this feedback and adjust the target audiences and the means to suit the opportunities. Use focus groups, surveys, informal discussions and suggestion schemes.

Energy champions

Once the energy awareness campaign has been developed and approved by senior management, it is useful to establish a network of volunteers to assist with the promotion of energy-saving values, messages and behaviour to colleagues. These energy champions can be very effective and a good resource. But they need to be carefully recruited, trained and equipped for their role. They also need ongoing support to stay effective and active.

Energy champions are particularly useful where energy use is spread over a large number of employees. For example, in a NHS trust, if the energy expenditure per year is divided by the total number of employees, the ratio is usually around £800 per employee per year. If every member of staff (say 3,000) does a little, it can make a big difference. In industry, the ratio might be £35,000 per employee per year, and this usually requires targeted training. In this situation, a few people can make a big difference in either wasting or saving energy.

Energy champions need to model the attitudes and behaviour they are seeking to promote. Energy champions need to lead by example, become the 'face' of the campaign to their colleagues, respond to questions, objections and problems, and act as the 'eyes and ears' of the campaign team.

The title 'energy champions' reflects the subject matter (energy) and what they are about (champions). The title needs to reflect their role in the campaign. If the campaign simply covers energy reduction, then 'energy' is fine. However, if the campaign also covers waste/recycling, travel, water and procurement, then terms like 'green', 'environmental' or 'sustainability' might be more appropriate. For some organizations, the term 'champion' might not fit into the culture. Other options could include 'representative', 'co-ordinator', 'facilitator', 'ambassador' or 'advocate'.

How many do we need?

First, you need to decide where you want them. Another application of the Pareto principle applies, where 80 per cent of the energy is used in 20 per cent of the largest buildings. If this is the ratio, then it is important to focus on the 20 per cent. For an estate with 100 buildings,

Appendix A People aspects/behavioural change

having at least one energy champion in each of the 20 largest buildings will be a priority. It is useful to start with a committed core group and gradually build up the numbers.

In a large, diverse estate such as a university or a hospital, it is helpful to have one energy champion per £40,000 of energy expenditure. So, for an annual £2 million energy bill, a good number is 50 energy champions. Also, as the campaign progresses, the geographical locations of energy champions can be mapped, and gaps identified and then targeted and filled with new energy champions. In an industrial setting, the focus is less likely to be on energy champions and more on the training of key managers and plant operatives who have hands-on control of significant energy use.

How do we recruit them?

It is best that employees volunteer for the role. Before volunteering they will need some indication of what is involved, and how much time it will involve. They may also be anxious about whether they will get their manager's approval and support.

At the University of St Andrews there is a network of environmental facilitators. In addition to the time required for training, the post is advertised as taking 30 minutes per week to do. However, these facilitators are alert to opportunities as they do their jobs, and give additional time of their own to evening walkabouts, when their areas are largely unoccupied.

Some staff will volunteer when a manager asks for volunteers, as the question implicitly suggests management support for the role. Another useful way of finding people quickly is to conduct an online environmental survey of the type described earlier, and to have a question at the end asking for staff to volunteer and send in their details.

What support do they need?

In addition to management support, energy champions need to know they have the support of the campaign team and their fellow energy champions.

Energy champions need training to equip them for their role. Often, training focuses on what they can do to make energy savings. However, what is often neglected is how to help them engage with, and influence, their colleagues. Understanding why colleagues behave the way they do, and how to influence their behaviour in a winning way, is vital. So the role is not only about technical solutions, but also about promoting behaviour change.

At a large, London acute NHS trust, the energy champions were given a special lanyard for their security passes, with the campaign logo and name printed thereon. The rest of the staff had to use the standard blue NHS lanyards so that the energy champions stood out.

Once energy champions get active, they will generate ideas. Some will be good housekeeping initiatives that can be implemented immediately. However, other opportunities will involve maintenance, changes in procedures and the use of controls, or low-cost investment.

It is important to register all ideas in a transparent way. Some organizations have specially designed apps to collect and categorize ideas from employees. There needs to be resources available to respond to these ideas, and it is important to reply personally to energy champions, to thank them and to inform them of the likely action that will be taken, and when.

Maintaining momentum

Once a campaign is up and running, a key challenge is to keep it going and fresh. It is inevitable that a campaign will have a finite life. In developing the campaign strategy plan it is useful to aim for a three-year programme. Much of the plan will focus on the first year. A review should take place at the end of the first year so that opportunities and priorities can be decided on for subsequent years. This allows for flexibility in planning a campaign, to consider the momentum element.

A helpful way to look at momentum is to distinguish actions or activities as being either 'special' or 'one-off', such as a launch event, or 'built-in', that is, integrated ways of operation that are permanent and continually in place.

Integration is the key to maintaining momentum. The fastest way to achieve integration is to analyse existing policies, procedures, management systems, methods of working, incentives and training to look for ways to integrate the sustainability message so that it becomes part of the DNA of the organization's culture. It simply becomes 'that's the way we do things around here'.

Some examples of integration are by incorporating energy in: induction training, job descriptions, staff appraisal, policies/procedures, energy management systems, environmental management systems, health and safety, quality, team meeting agendas, publications, staff training, procurement, maintenance and contractors.

Appendix B List of complementary standards

Table B1: British standards

Identifier	Title
BS 845-1	Methods for assessing thermal performance of boilers for steam, hot water and high temperature heat transfer fluids — Part 1: Concise procedure
BS 5225-1	Photometric data for luminaires — Part 1: Photometric measurements
BS 5422	Method for specifying thermal insulating materials for pipes, tanks, vessels, ductwork and equipment operating within the temperature range -40 °C to +700 °C
BS 5991	Specification for indirect gas fired forced convection air heaters with rated heat inputs greater than 330 kW but not exceeding 2 MW for industrial and commercial space heating — Safety and performance requirements (excluding electrical requirements) (2nd family gases)
BS 7913	Guide to the conservation of historic buildings

Table B2: European standards

Identifier	Title
EN 267	Automatic forced draught burners for liquid fuels
EN 308	Heat exchangers — Test procedures for establishing performance of air to air and flue gases heat recovery devices
EN 378-1	Refrigerating systems and heat pumps — Safety and environmental requirements — Part 1: Basic

Appendix B List of complementary standards

Identifier	Title
	requirements, definitions, classification and selection criteria
EN 416-2	Single burner gas-fired overhead radiant tube heaters for non-domestic use — Part 2: Rational use of energy
EN 419-2	Non-domestic gas-fired overhead luminous radiant heaters — Part 2: Rational use of energy
EN 621	Non-domestic gas-fired forced convection air heaters for space heating not exceeding a net heat input of 300 kW, without a fan to assist transportation of combustion air and/or combustion products
EN 676	Automatic forced draught burners for gaseous fuels
EN 810	Dehumidifiers with electrically driven compressors — Rating tests, marking, operational requirements and technical data sheet
EN 1020	Non-domestic forced convection gas-fired air heaters for space heating not exceeding a net heat input of 300 kW incorporating a fan to assist transportation of combustion air or combustion products
EN 12599	Ventilation for buildings — Test procedures and measurement methods to hand over air conditioning and ventilation systems
EN 12792	Ventilation for buildings – Symbols, terminology and graphical symbols
EN 12900	Refrigerant compressors – Rating conditions, tolerances and presentation of manufacturer's performance data
EN 12952-15	Water-tube boilers and auxiliary installations — Part 15: Acceptance tests
EN 12953-11	Shell boilers — Part 11: Acceptance tests
EN 13032-1	Light and lighting — Measurement and presentation of photometric data of lamps and luminaires — Part 1: Measurement and file format

Energy Audits

Appendix B List of complementary standards

Identifier	Title
EN 13187	Thermal performance of buildings — Qualitative detection of thermal irregularities in building envelopes — Infrared method (ISO 6781:1983 modified)
EN 13215	Condensing units for refrigeration — Rating conditions, tolerances and presentation of manufacturer's performance data
EN 13240	Roomheaters fired by solid fuel — Requirements and test methods
EN 13363	Solar protection devices combined with glazing — Calculation of solar and light transmittance
EN 13771	Compressors and condensing units for refrigeration — Performance testing and test methods
EN 13779	Ventilation for non-residential buildings — Performance requirements for ventilation and room-conditioning systems
EN 13829	Thermal performance of buildings — Determination of air permeability of buildings — Fan pressurization method (ISO 9972:1996, modified)
EN 13842	Oil fired forced convection air heaters — Stationary and transportable for space heating
EN 14511	Air conditioners, liquid chilling packages and heat pumps with electrically driven compressors for space heating and cooling
EN 14785	Residential space heating appliances fired by wood pellets — Requirements and test methods
EN 15193	Energy performance of buildings — Energy requirements for lighting
EN 15217	Energy performance of buildings — Methods for expressing energy performance and for energy certification of buildings
EN 15232	Energy performance of buildings — Impact of building automation, controls and building management

Appendix B List of complementary standards

Identifier	Title
EN 15239	Ventilation for buildings — Energy performance of buildings — Guidelines for inspection of ventilation systems
EN 15240	Ventilation for buildings — Energy performance of buildings — Guidelines for inspection of air-conditioning systems
EN 15241	Ventilation for buildings — Calculation methods for energy losses due to ventilation and infiltration in commercial buildings
EN 15242	Ventilation for buildings — Calculation methods for the determination of air flow rates in buildings including infiltration
EN 15243	Ventilation for buildings — Calculation of room temperatures and of load and energy for buildings with room conditioning systems
EN 15251	Indoor environmental input parameters for design and assessment of energy performance of buildings addressing indoor air quality, thermal environment, lighting and acoustics
EN 15255	Energy performance of buildings — Sensible room cooling load calculation — General criteria and validation procedures
EN 15265	Energy performance of buildings — Calculation of energy needs for space heating and cooling using dynamic methods — General criteria and validation procedures
EN 15316	Heating systems in buildings — Method for calculation of system energy requirements and system efficiencies
EN 15377	Heating systems in buildings — Design of embedded water based surface heating and cooling systems
EN 15378	Heating systems in buildings — Inspection of boilers and heating systems

Appendix B List of complementary standards

Identifier	Title
EN 15459	*Energy performance of buildings — Economic evaluation procedure for energy systems in buildings*
EN 15603	*Energy performance of buildings — Overall energy use and definition of energy ratings*
EN 16212	*Energy efficiency and savings calculation, top-down and bottom-up methods*
EN 16231	*Energy efficiency benchmarking methodology*
EN 16247-1	*Energy audits — Part 1: General requirements*
EN 60034-1	*Rotating electrical machines — Part 1: Rating and performance*
EN 60034-2-1	*Rotating electrical machines — Part 2-1: Standard methods for determining losses and efficiency from tests (excluding machines for traction vehicles)*

Table B3: International standards

Identifier	Title
ISO 6946	*Building components and building elements — Thermal resistance and thermal transmittance — Calculation method*
ISO 7345	*Thermal insulation — Physical quantities and definitions*
ISO 9251	*Thermal insulation — Heat transfer conditions and properties of materials — Vocabulary*
ISO 9288	*Thermal insulation — Heat transfer by radiation — Physical quantities and definitions*
ISO 10077	*Thermal performance of windows, doors and shutters — Calculation of thermal transmittance*
ISO 10211	*Thermal bridges in building construction — Heat flows and surface temperatures — Detailed calculations*

Appendix B List of complementary standards

Identifier	Title
ISO 10456	Building materials and products — Hygrothermal properties — Tabulated design values and procedures for determining declared and design thermal values
ISO 11011	Compressed air — Energy efficiency — Assessment
ISO 12569	Thermal performance of buildings and materials — Determination of specific airflow rate in buildings — Tracer gas dilution method
ISO 12655	Energy performance of buildings — Presentation of measured energy use of buildings
ISO 13153	Framework of the design process for energy-saving single-family residential and small commercial buildings
ISO 13370	Thermal performance of buildings — Heat transfer via the ground — Calculation methods
ISO 13579-1	Industrial furnaces and associated processing equipment — Method of measuring energy balance and calculating efficiency — Part 1: General methodology
ISO 13579-2	Industrial furnaces and associated processing equipment — Method of measuring energy balance and calculating efficiency — Part 2: Reheating furnaces for steel
ISO 13602	Technical energy systems — Methods for analysis
ISO 13786	Thermal performance of building components — Dynamic thermal characteristics — Calculation methods
ISO 13789	Thermal performance of buildings — Transmission and ventilation heat transfer coefficients — Calculation method
ISO 13790	Energy performance of buildings — Calculation of energy use for space heating and cooling
ISO 13791	Thermal performance of buildings — Calculation of internal temperatures of a room in summer without

Appendix B List of complementary standards

Identifier	Title
	mechanical cooling — General criteria and validation procedures
ISO 13792	Thermal performance of buildings — Calculation of internal temperatures of a room in summer without mechanical cooling — Simplified methods
ISO 14044	Environmental management — Life cycle assessment — Requirements and guidelines
ISO 14414[13]	Pump system energy assessment
ISO 14683	Thermal bridges in building construction — Linear thermal transmittance — Simplified methods and default values
ISO 15927	Hygrothermal performance of buildings — Calculation and presentation of climatic data
ISO 16346	Energy performance of buildings — Assessment of overall energy performance
ISO 16484-7[14]	Building automation and control systems (BACS) — Part 7: The contribution of BACS to energy performance of buildings
ISO 16818	Building environment design — Energy efficiency — Terminology
ISO 19011	Guidelines for auditing management systems
ISO 20140-1	Automation systems and integration — Evaluating energy efficiency and other factors of manufacturing systems that influence the environment — Part 1: Overview and general principles
ISO 23045	Building environment design — Guidelines to assess energy efficiency of new buildings
ISO 25745-1	Energy performance of lifts, escalators and moving walks — Part 1: Energy measurement and verification

---

[13] ISO 14414 is due to be published in 2014.
[14] ISO 16484-7 is due to be published in 2014.

Appendix B List of complementary standards

Identifier	Title
ISO 50001	*Energy management systems — Requirements with guidance for use*
ISO/IEC 17021	*Conformity assessment — Requirements for bodies providing audit and certification of management systems*

Further Reading

Doty, S and Turner, WC (2012) Energy Management Handbook, Eighth Edition, Lilburn: Fairmont Press

Fawkes, S (2013) *Energy Efficiency: The Definitive Guide to the Cheapest, Cleanest, Fastest Source of Energy*, Farnham: Gower

Mulholland, J (forthcoming) *Greening the Workforce: Energy Programmes and Employee Behaviour*, Farnham: Gower

Oung, K (2013) *Energy Management in Business: The Manager's Guide to Maximising and Sustaining Energy Reduction*, Farnham: Gower

Vesma, V (2011) *Energy Management Principles and Practice*, Second edition, London: British Standards Institution (BIP 2187)

EN 16247, *Energy audits – Part 3: Processes*

ISO 9001, *Quality management systems – Requirements*

ISO 14001, *Environmental management systems – Requirements with guidance for use*

ISO 50002, *Energy audits – Requirements with guidance for use*

OHSAS 18001, *Occupational health and safety management systems – Requirements*

The Chartered Institution of Building Services Engineers (CIBSE), *Guide A: Environmental Design*

American National Standards Institute (ANSI)/American Society of Heating, Refrigerating and Air-Conditioning Engineers (ASHRAE), *Standard 62.1, Ventilation for Acceptable Indoor Air Quality*

References

[1] MacKay, David JC (2009) *Sustainable Energy: Without the Hot Air*, Cambridge: UIT

[2] Bressand, F, Farrell, D, Haas, P, et al. (2007) *Curbing Global Energy Demand Growth: The Energy Productivity Opportunity*, McKinsey Global Institute, May 2007, And Hartmann, A, Farrell, D, Graubner, M and Remes, J (2008) *Capturing the European Energy Productivity Opportunity*, McKinsey Global Institute, September 2008

[3] Dobbs, R, Oppenheim, J, Thompson, F, et al. (2011) *Resource Revolution: Meeting the World's Energy, Materials, Food, and Water Needs*, McKinsey Global Institute, November 2011

[4] MacKay, David JC (2009) *Sustainable Energy: Without the Hot Air*, Cambridge: UIT

[5] Rigby, D and Bilodeau, B (2013) *Management Tools & Trends 2013*, Bain & Company

[6] Werbach, A (2009) *Strategy for Sustainability: A Business Manifesto*, Boston: Harvard Business Press

[7] Rockström, J, Steffen, W, Noone, K, et. al. (2009) 'A safe operating space for humanity', *Nature*, Vol. 461, 24 September 2009, pp. 472–475

[8] Ochoa, P (2013) *Electricity Capacity Assessment Report 2013*, Report to the Secretary of State, 105/13, 27 June 2013, Ofgem

[9] Chazan, G (2013) *'Invisible Fuel' Promises More Secure Future'*, Modern Energy, A FT Special Report, Tuesday, 3 June 2013, London: The Financial Times Ltd

[10] Enkvist, P, Naucler, T and Riese, J (2008) 'What countries can do about cutting carbon emissions' *McKinsey Quarterly*, April 2008

[11] Esty, DC, and Winston, A (2009) *Green to Gold: How Smart Companies Use Environmental Strategy to Innovate, Create Value, and Build Competitive Advantage*. New Jersey: John Wiley & Sons. Also in Harvard Business Review (2013) *The New Global Energy Economy: Towards a New Future*. A report by Harvard Business Review Analytic Services, Boston: Harvard Business Review

References

[12] Groysberg, B, Healy, P, Nohria, N, and Serafeim, G (2012) 'What makes analysts say "buy"?', *Harvard Business Review*, November 2012, Boston: Harvard Business Review Press

[13] Edelman (2012) *2012 Edelman goodpurpose® study: Executive Summary*

[14] Courtice, P (2013) 'The critical link: strategy and sustainability in leadership development', *The State of Sustainability Leadership*, Cambridge: University of Cambridge Programme for Sustainability Leadership Publication

[15] Edelman (2012) *2012 Edelman goodpurpose® study: Executive Summary*

[16] PricewaterhouseCoopers (2012) *PwC Millennials at Work: Reshaping the workplace in Financial Services*, PwC

[17] Dougherty, A, Mitchell-Jackson, J and Wellner, P (2010) *Ethnographic Inquiry in Energy: Exploring Meaning-Making and Sociality in Language Use, Program Participation, and Behavioral Choice*, 2010 ACEEE Summer Study on Energy Efficiency in Buildings

[18] Murray, S [Holloway, N and Thody, J (eds)] (2011) *Unlocking the Benefits of Energy Efficiency: An Executive Dilemma*, Economist Intelligence Unit report, London: The Economist

[19] Dougherty, K and Mikytuck, H, *GE's eco Treasure Hunt Checklist*, GE Capital

[20] Winston, A (2013) 'The Inside Story of Diageo's Stunning Carbon Achievement', Harvard Business Review Blog Network, 20 February 2013. Available at: http://blogs.hbr.org/winston/2013/02/the-inside-story-of-diageos-st.html [accessed 6 May 2013]

[21] Beavis, S (2011) 'The Environment Agency – simple rules, clever kit', Sustainability case study, *The Guardian*, 26 May 2011. Available at: www.guardian.co.uk/sustainable-business/environment-agency-simple-rules-clever-kit [accessed 19 June 2013]

[22] World Economic Forum (2009) *Supply Chain Decarbonization*, Geneva: World Economic Forum

[23] Lavery, G, Pennell, N, Brown, S and Evans, S (2013) *The Next Manufacturing Revolution: Non-Labour Resource Productivity and its Potential for UK Manufacturing*, July 2013, Lavery Pennell, 2degrees and The Institute for Manufacturing

References

[24] Hodder, S (2010) 'Thinking about energy', *Ethical Performance*, Autumn 2010. Available at: www.ethicalperformance.com/bestpractice/casestudy/96 [accessed 6 July 2013]

[25] Vogelaar, R (2010) 'New departure procedure for Airbus A380 Singapore Airlines at London–Heatrow[sic]', Aviationnews.eu, 2 March 2010. Available at: www.aviationnews.eu/2010/03/02/new-departure-procedure-for-airbus-a380-singapore-airlines-at-london-heatrow [accessed 21 July 2013]

[26] 'London Heathrow Airport', Wikipedia. Available at: http://en.wikipedia.org/wiki/London_Heathrow_Airport [accessed 6 July 2013]

[27] Kingsley-Jones, M (2010) 'Airbus details A380's new, more efficient Heathrow departure procedures', *Flightglobal*, 4 March 2010. Available at: www.flightglobal.com/news/articles/airbus-details-a380s-new-more-efficient-heathrow-departure-procedures-339067 [accessed 21 July 2013]

[28] Heathrow Airport, 'Airbus A380 new departure procedures'. Sustainability case study. Available at: www.heathrowairport.com/about-us/community-and-environment/sustainability/case-studies/airbus-a380-new-departure-procedures [accessed 6 July 2013]

[29] Goldberg, A, Holdaway, E, Reinaud, J and O'Keeffe, S (2012) *Promoting Energy Savings and GHG Mitigation through Industrial Supply Chain Initiatives*, May 2012, Institute for Industrial Productivity

[30] Chanel, G (2011) 'High yields from sustainable supplies', *T Magazine*, Issue 5, pp. 26-29

[31] Murray, S, [Holloway, N and Thody, J (eds)] (2011) *Unlocking the Benefits of Energy Efficiency: An Executive Dilemma*, Economist Intelligence Unit report, London: The Economist. CEMEX (2012) CEMEX's position on climate change. 1st July 2013. Available at: www.cemex.com/MediaCenter/Files/CEMEX_POSITION_on_Climate_Change.pdf [accessed: 17 June 2014]

[32] Winston, A (2013) 'The inside story of Diageo's stunning Carbon achievement', Harvard Business Review Blog Network, 20 February 2013. Available at: http://blogs.hbr.org/winston/2013/02/the-inside-story-of-diageos-st.html [accessed 6 May 2013]

[33] Norton, DP (2007) 'Strategy Execution: A Competency that Creates Competitive Advantage', A Palladium White Paper, Palladium Group

References

[34] Hodder, S (2010) 'Thinking about energy', *Ethical Performance*, Autumn 2010. Available at: www.ethicalperformance.com/bestpractice/casestudy/96 [accessed 6 July 2013]

[35] Kaplan, RS and Norton, DP (2008) 'Mastering the management system', *Harvard Business Review*, January 2008, Boston: Harvard Business Review Press

[36] Reyna, E, Hiller, J, Riso, C and Jay, J (2012) *The Virtuous Cycle Of Organizational Energy Efficiency: A Fresh Approach to Dismantling Barriers*, 2012 ACEEE Summer Study on Energy Efficiency in Buildings, pp.10-295–10-309

[37] ISO 50001:2011, *Energy management systems — Requirements with guidance for use*, Clause 3.9

[38] Kaplan, R and Norton, D (2001) *The Strategy-focused Organization: How Balanced Scorecard Companies Thrive in the New Business Environment*, Boston: Harvard Business School Press. Kaplan, R and Norton, D (2008) *The Execution Premium: Linking Strategy to Operations for Competitive Advantage*, Boston: Harvard Business Press

[39] Kotter, JP (1996) *Leading Change*. Boston: Harvard Business School Press

[40] Keller, S and Price, C (2011) *Beyond Performance: How Great Organizations Build Ultimate Competitive Advantage*, New Jersey: John Wiley & Sons

[41] Aflaki, S and Kleindorfer, PP (2010) *Going Green: The Pfizer Freiburg Energy Initiative*, INSEAD Social Innovation Centre, 03/2010-5680, Fontainebleau: INSEAD

[42] Chachoua, E and Hulse, J (2013) *Investing in Energy Efficiency in Europe's buildings: A View from the Construction and Real Estate Sectors*, Economist Intelligence Unit report, The Economist

[43] Molloy, C (2013) 'Matthews Coach Hire'. Available at: http:aems.ie/Pages/Matthews.aspx [accessed 10 June 2013]

[44] US Department of Energy (2010) *Success Story: Ingersoll Rand Discovers Hidden Savings with a Three-Tiered Energy Audit Model*, October 2010, A report by US Department of Energy Industrial Technologies Program

[45] Verdantix (2009) *M&S Brand Benefits From Its Climate Plan*, 15 March 2009, Verdantix

[46] Bara, G (2010) *Mandate to CEN, CENELEC and ETSI for Elaboration of Standards Regarding Energy Audits*, M/479 EN, 13 December 2010, Brussels: European Commission, Directorate-General for Energy

[47] ISO 50002:2014, *Energy audits — Requirements with guidance for use*

[48] For further details of the energy audit types, please see Annex A of ISO 50002.

[49] Baczko, K, Malkin, AE, Campbell, I, et al. *A Landmark Sustainability Program for the Empire State Building: A Model for Optimizing Energy Efficiency, Sustainable Practices, Operating Expenses and Long Term Value in Existing Buildings.* A report by Jones Lang LaSalle, Clinton Climate Initiative, Rocky Mountain Institute and Johnson Controls, Inc.

[50] Shaw, J (2012) 'A green empire: How Anthony Malkin'84 engineered the largest "green" retrofit ever', March–April 2012, *Harvard Magazine*

[51] Shaw, J (2012) 'Green Engineering', March–April 2012, *Harvard Magazine*

[52] Dunn, M (2013) 'GMS Irvine – Energy Efficiency', A presentation in The European Chemical Industry Council's Responsible Care and SPICE3 Conference 2013: 'Small Improvement, Great Effects', 2–4 December 2013, Amsterdam. Available at: www.cefic.org/Documents/ResponsibleCare/RC-SPiCE3-Conference-2013/Day1-05-Sustainability-Programme-M-Dunn-GSK.pdf [accessed 15 April 2014] and in CMR Consultants' Energy Kaizen: Accelerated Energy and CO_2 Savings Programme. Available at: www.energ-group.com/media/570098/energy_kaizen_brochure.pdf [accessed 15 April 2014]

[53] Murray, S [Gardner, B and Tabary, Z (eds)] (2013) *Sustainability Insights: Learning from Business Leaders*, A Coca-Cola Enterprises report, written by the Economist Intelligence Unit, 1 October 2013

[54] Coca-Cola Hellenic Bottling Company (Coca-Cola HBC) (2010) 'Energy and climate change'. Available at: www.coca-colahellenic.com/sustainability/environment/energyandclimate/heat_and_power_plant [accessed 8 May 2013]

[55] Economist Intelligence Unit (2013) *Global Manufacturing Outlook – Competitive advantage: enhancing supply chain networks for efficiency and innovation*, An Economist Intelligence Unit research program sponsored by KPMG International.

Economist Intelligence Unit (2012) *Global Manufacturing Outlook: Fostering Growth through Innovation*, Global research commissioned by KPMG International from the Economist Intelligence Unit.

References

Economist Intelligence Unit (2011) *Global Manufacturing Outlook: Growth while Managing Volatility*, Global research commissioned by KPMG International from the Economist Intelligence Unit

[56] Crawford Brown, D (2013) Driving down carbon through the supply chain, University of Cambridge Programme for Sustainability Leadership

[57] GlaxoSmithKline (GSK) (2013) *GSK Corporate Responsibility Report 2012*, Brentford: GSK

[58] Tesco Press Office (2013) 'Tesco launches buying club to help suppliers to cut carbon and energy use', press release, 27 September 2013, Tesco

[59] Harvard Business Review Analytic Services (2013) *The Future of Energy*, A Harvard Business Review Analytic Services Report

[60] Rigby, D and Bilodeau, B (2013) *Management Tools &Trends 2013*, Bain and Company

[61] Chazan, G (2013) *'Invisible Fuel' Promises More Secure Future*, Modern Energy, A FT Special Report, Tuesday, 4 June 2013, London: The Financial Times Ltd

[62] Hansen, G (2012) *Facebook's Open Compute Project: An Industry Wide Collaborative Initiative to Accelerate Enterprise Datacenter Energy Efficiency*, 2012 ACEEE Summer Study on Energy Efficiency in Buildings

[63] Facebook (2011) 'Open compute project'. Available at: http://opencompute.org [accessed 18 April 2011]

[64] Lockwood, C (2006) 'Building the green way', *Harvard Business Review*, June 2006, Boston: Harvard Business Review Press

[65] Carris, J and Brunelli, G (2011) The Velodrome, the most energy efficient venue on the Olympic Park, Learning legacy: Lessons learned from the London 2012 Games construction project, ODA 2010/374, October 2011

[66] Sustainable Energy Ireland (2009) 'Case Study: Pfizer', Energy Efficient Design Special Working Group, EED SWG 006, Q1 2009, Dublin: Sustainable Energy Ireland

[67] 2degrees (2014) *Overcoming the Engagement Barrier: 2degrees Sustainable Business Trends Tracker*, Executive Summary

[68] Broad, L (2013) 'Dawn Meats uses space technology to achieve thermal energy reduction', 2degrees, Awards Entry, 1 February 2013. Available at:.

www.2degreesnetwork.com/groups/2degrees-community/resources/dawn-meats-uses-space-technology-achieve-thermal-energy-reduction/ [accessed 30 December 2013]

[69] Werbach, A (2009) *Strategy for Sustainability: A Business Manifesto*, Boston: Harvard Business Press

[70] Unilever (2012) 'Sustainable living': 'water use by consumers'. Available at: www.unilever.com/sustainable-living-2014/reducing-environmental-impact/water-use/water-use-by-consumers/? [accessed 17 June 2014]

[71] Sharpe, L (2014) 'London project puts waste heat to good use', *E & T (Engineering and Technology Magazine)*, January 2014, Vol. 8, Issue 12, p.12, The Institution of Engineering and Technology. And in Thrower, M (ed.) (2014) 'Northern Line coming to 500 London homes', *Energy in Buildings and Industry*, January 2014, p.9

[72] European Union (2012) Directive 2012/27/EU of the European Parliament and of the Council of 25 October 2012 on energy efficiency, amending Directives 2009/125/EC and 2010/30/EU and repealing Directives 2004/8/EC and 2006/32/EC, OJ L 315, 14.11.2012, pp.1–56 (the Energy Efficiency Directive)

[73] Department of Energy and Climate Change. (2014) Energy Saving Opportunity Scheme (ESOS): Guide to ESOS. Version 1. June 2014.

[74] Cormack, H (2014) *Energy Saving Opportunity Scheme: Impact Assessment*. IA No. DECC0142. Dated 24 June 2014.

[75] Kennett, S (2013) 'RBS Group uses energy efficiency to cut £1.2 million from its annual energy costs', 2degrees, Case Study, 17 July 2013. Available at: www.2degreesnetwork.com/groups/energy-carbon-management/resources/rbs-group-uses-energy-efficiency-cut-12-million-its-annual-energy-costs [accessed 30 December 2013]

[76] Van Zyl, R (2013) Making Energy Savings a Reality. Saldanha Works – A Case Study. Presentation at African Utility Week: Delivering Beyond Tomorrow.

[77] Department of Trade and Industry Republic of South Africa, Department of Energy Republic of South Africa, and United Nations Industrial Development Organisation (UNIDO), et.al., *Introduction and Implementation of an Energy Management System and Energy Systems Optimization. Case Study: ArcelorMittal Saldanha Works*.

[78] Department for Business, Innovation & Skills (2012) 'Business population estimates for the UK and regions 2012', 17 October 2012. Available at:

References

www.gov.uk/government/uploads/system/uploads/attachment_data/file/80247/bpe-2012-stats-release-4.pdf [accessed: 17 June 2012]

[79] Kannan, R and Boie, W (2003) 'Energy management practices in SME—case study of a bakery in Germany' *Energy Conversion and Management*, Vol. 44, Iss. 6, April 2003, pp.945–959

[80] Department of Energy and Climate Change (2014) *Energy Saving Opportunity Scheme (ESOS): Guide to ESOS*. Version 1. June 2014.

[81] Murray, S [Gardner, B and Tabary, Z (eds)] (2013) *Sustainability Insights: Learning from Business Leaders*, A Coca-Cola Enterprises report, written by the Economist Intelligence Unit, 1 October 2013

[82] European Commission (2014) 'Improving corporate governance: Europe's largest companies will have to be more transparent about how they operate', European Commission Statement 14/124.15/04/2014, Brussels. And European Commission (2014). 'Disclosure of non-financial and diversity information by large companies and groups - Frequently asked questions' European Commission Memo 14/301. 15/04/2014. Brussels.

[83] Eccles, RG, Cheng, B and Thyne, S (2010) *'Southwest Airlines One Report™'*, *Harvard Business School Case 9-411-042*, September 2010, Boston: Harvard Business School Publishing

[84] Groysberg, B, Healy, P, Nohria, N and Serafeim, G (2012) 'What makes analysts say 'buy'?', *Harvard Business Review*, November 2012, Boston: Harvard Business School Publishing

[85] Chatterji, AK and Toffel, MW (2010) 'How firms respond to being rated', *Strategic Management Journal*, Vol. 31, 9, September 2010, pp.917–945, John Wiley & Sons. Doshi, AR, Dowell, GWS and Toffel, MW (2013) 'How firms respond to mandatory information disclosure', *Strategic Management Journal*, Vol. 34,. 10, pp.1209–1231, October 2013, John Wiley & Sons

[86] Walmart 'Greenhouse gas emissions'. Available at: http://corporate.walmart.com/global-responsibility/environment-sustainability/greenhouse-gas-emissions [accessed 27 August 2013]

[87] Winston, A (2012) 'How Walmart's green performance reviews could change retail for good', *Harvard Business Review Blog Network*, 2 October 2012. Available at: blogs.hbr.org/Winston/2012/10/how-walmarts-green-performance.html [accessed 27 August 2013]

[88] Oxford City Council (2010) 'Low Carbon Oxford launched', press release, 14 October 2010, Oxford City Council

[89] Oxford City Council (2014) 'Low Carbon Oxford'. Available at: www.oxford.gov.uk/lowcarbonoxford [accessed 17 June 2014]

[90] ISO Focus (2013) 'Energy Boooooost for Costa Coffee', *ISOfocus*,. 101, November–December 2013, pp.46–47

[91] Kennett, S (2013) 'Refurbishment of PwC HQ sets benchmark for sustainable building', 2degrees, 20 December 2013. Available at: www.2degreesnetwork.com/groups/2degrees-community/resources/refurbishment-pwc-hq-sets-benchmark-sustainable-building [accessed 20 December 2013]

[92] Newham University Hospital – Barts Health NHS Trust (2010) 'Newham University Hospital RE:FIT Information Pack', September 2010

[93] Turnbull Henry, E (2013) 'How the adidas Group prioritizes investments in energy efficiency', 2degrees, Case Study, 3 June 2013. Available at: www.2degreesnetwork.com/groups/energy-carbon-management/resources/how-adidas-group-prioritizes-investments-energy-efficiency [accessed 20 December 2013]

[94] Jones Lang LeSalle (2012) *A Tale of Two Buildings: Are EPCs a true indicator of energy efficiency?* A report by Jones Lang LaSalle and Better Buildings Partnership

[95] TSO (The Stationery Office) (2010) *The Building Regulations 2010: Approved Document L1A, Conservation of Fuel and Power in New Dwellings*, London: TSO (The Stationery Office)

[96] TSO (The Stationery Office) (2010) *The Building Regulations 2010: Approved Document L1B, Conservation of Fuel and Power in Existing Dwellings*, London: TSO (The Stationery Office)

[97] TSO (The Stationery Office) (2010) *The Building Regulations 2010: Approved Document L2A, Conservation of Fuel and Power in New Buildings Other Than Dwellings*, London: TSO (The Stationery Office)

[98] TSO (The Stationery Office) (2010) *The Building Regulations 2010: Approved Document L2B, Conservation of Fuel and Power in Existing Buildings Other Than Dwellings*, London: TSO (The Stationery Office)

[99] Passivhaus, 'The Passivhaus Standard'. Available at: www.passivhaus.org.uk/standard.jsp?id=122 [accessed 15 May 2014]

[100] Passivhaus, 'EnerPHit Standard'. Available at: www.passivhaus.org.uk/page.jsp?id=20 [accessed 15 May 2014]

[101] Great Britain Department of Health – Gateway Review Estates and Facilities Division (2007) *Heating and Ventilation Systems – Health*

References

Technical Memorandum 03-01: Specialised Ventilation for Healthcare Premises – Part A: Design and Validation, London: TSO (The Stationery Office); and
Great Britain Department of Health – Gateway Review Estates and Facilities Division (2007) Heating and Ventilation Systems – Health Technical Memorandum 03-01: Specialised Ventilation for Healthcare Premises – Part B: Operational Management and Performance Verification, London: TSO (The Stationery Office)

[102] Department of Energy and Climate Change and HM Revenue and Customs (2013), Energy technology criteria list, 1 July 2013. Available at: https://etl.decc.gov.uk/etl/site/criteria.html [accessed 17 July 2014]

[103] American Society of Heating, Refrigerating and Air-Conditioning Engineers. (2012) *Thermal Guidelines for Data Processing Environments: ASHRAE Datacom Series*. Third Edition.

[104] Bowes-Phipps, S (2011) 'Reduction and Re-use of Energy in Institutional Data Centers (RARE-IDC), 2011 Green Enterprise IT Award Winner, Innovation in a Smaller Data Center < 1,000 sq ft, University of Hertfordshire', Uptime Institute Symposium 2011, 9–12 May 2011, Santa Clara, California

[105] Savage, SL (2009) *The Flaw of Averages: Why We Underestimate Risk in the Face of Uncertainty*. New Jersey: John Wiley & Sons

[106] Hoshide, RK (1994) 'Electric motor do's and don'ts', *Energy Engineering*, Vol. 91, No. 1, pp. 6-24

[107] Kennett, S (2013) 'RBS Group uses energy efficiency to cut £1.2 million from its annual energy costs', 2degrees, Case Study, 17 July 2013. Available at: www.2degreesnetwork.com/groups/energy-carbon-management/resources/rbs-group-uses-energy-efficiency-cut-12-million-its-annual-energy-costs [accessed 30 December 2013]

[108] Clayton, J (2003) 'Writing an Executive Summary That Means Business', Harvard Management Communication Letter, July 2003, Boston: Harvard Business School Publishing

[109] de Rond, M (2012) *There Is an I in Team: What Elite Athletes and Coaches Really Know about High Performance*, Boston: Harvard Business Review Press

[110] Currie, M (2013) 'Communication for Consultants', IChemE Consultancy Special Interest Group webinar

[111] Ferguson, M (2013) 'Job titles aren't that important', Harvard Business Review Blog Network, 24 April 2013. Available at: blogs.hbr.org/cs/2013/04/don't_filter_job_candidates_by.html [accessed 29 April 2013]

[112] Currie, M (2013) 'Consultancy: the rise of the independent expert', in The Chemical Engineer (TCE) supplement: 'Consultants and Contractors File 2013', *The Chemical Engineer*, Issue 865, July 2013, Rugby: Institution of Chemical Engineers

[113] Christensen, CM, Wang, D and van Bever, D (2013) 'Consulting on the cusp of disruption' *Harvard Business Review*, October 2013, Harvard Business Review Press

[114] PAS 51215:2014, *Energy efficiency assessment – Competence of a lead energy assessor – Specification*

[115] de Rond, M (2012) *There Is an I in Team: What Elite Athletes and Coaches Really Know about High Performance*, Boston: Harvard Business Review Press

[116] Lencioni, P. (2010*) Getting Naked: A Business Fable About Shedding the Three Fears That Sabotage Client Loyalty*. San Francisco: Jossey-Bass

[117] Private communication with EDF Climate Corps representative

[118] Casciaro, T and Sousa Lobo, M (2005) 'Competent jerks, lovable fools, and the formation of social networks', *Harvard Business Review*, June 2005, Boston: Harvard Business Review Press. de Rond, M (2012) *There Is an I in Team: What Elite Athletes and Coaches Really Know about High Performance*. Boston: Harvard Business Review Press

[119] Energy Efficient Buildings Hub (2012) *Instrumenting Navy Yard Building 101*, A US Department of Energy, Energy Innovation Hub Report. Energy Efficient Buildings Hub (2012) *Variation in Energy Audits: A Case Study of Navy Yard Building 101*, A US Department of Energy, Energy Innovation Hub Report

[120] The Dow Chemical Company (2013) *The Dow Chemical Company 2012 Global Reporting Initiative (GRI) Report*. Available at: www.dow.com/sustainability/pdf/35865-2012%20Sustainability%20Report.pdf [accessed 18 November 2013]. And US Department of Energy (2007) *Dow Chemical Company: Assessment Leads to Steam System Energy Savings in a Petrochemical Plant*. Available at: www1.eere.energy.gov/manufacturing/tech_assistance/pdfs/42009.pdf [accessed 18 November 2013]

[121] Walmart (2013) 'Cutting our energy use through innovation'. Available at:

References

http://corporate.walmart.com/global-responsibility/environment-sustainability/buildings [accessed 18 November 2013]

[122] Sainsbury's (2013) 'Energy efficiency & store generation technologies', presented by Phil Osborn at 5th Smart Grids & Cleanpower Conference, 5 June 2013, Cambridge. Available at: www.cir-strategy.com/uploads/sgcp13osborn.pdf [accessed 18 November 2013]

[123] Telstra (2013) 'Environmental impact: reducing our environmental impact'. Available at: www.telstra.com.au/uberprod/groups/webcontent/@corporate/@about/documents/document/uberstaging_244378.pdf [accessed 18 November 2013]

[124] Stena Line (2013) 'Reducing the energy consumption'. Available at: www.stenaline.com/en/stena-line/corporate/environment/energy [accessed 18 November 2013]. And Hong Liang, L (2013) 'Stena Bulk achieves $9.7m in savings from energy efficiency', *Seatrade Global*, 28 June 2013. Available at: www.seatrade-global.com/news/europe/stena-bulk-achieving-$97m-savings-through-energy-efficiency.html [accessed 18 November 2013]